Viral Cytopathology

Authors

Hubert H. Malherbe
Department of Microbiology
University of Texas Health Science Center
San Antonio, Texas

Margaret Strickland-Cholmley
Department of Microbiology
University of Texas Health Science Center
San Antonio, Texas

CRC Press
Taylor & Francis Group
Boca Raton London New York

CRC Press is an imprint of the
Taylor & Francis Group, an **informa** business

First published 1980 by CRC Press
Taylor & Francis Group
6000 Broken Sound Parkway NW, Suite 300
Boca Raton, FL 33487-2742

Reissued 2018 by CRC Press

Library of Congress Cataloging in Publication Data

Main entry under title:

Malherbe, Hubert H.
Viral Cytophathology

Includes bibliographical indexes.
1. Virus diseases — Diagnosis. 2. Pathology, Cellular. 3. Viruses — Identification. 4. Stains and staining (Microscopy). I. Strickland-Cholmley, Margaret, joint author. II. Title [DNLM: 1. Cell transfomration, Viral. 2. Virus diseases — Pathology. QW160 M249y]
RC 114.5.M35 5567-2 80-11228
ISBN 0-8493-616,9'2507582

A Library of Congress record exists under LC control number: 80011228

Publisher's Note
The publisher has gone to great lengths to ensure the quality of this reprint but points out that some imperfections in the original copies may be apparent.

Disclaimer
The publisher has made every effort to trace copyright holders and welcomes correspondence from those they have been unable to contact.

ISBN 13: 978-1-315-89844-5 (hbk)
ISBN 13: 978-1-351-07754-5 (ebk)

Visit the Taylor & Francis Web site at http://www.taylorandfrancis.com and the
CRC Press Web site at http://www.crcpress.com

PREFACE

The need for a means of providing both evidence of viral activity and a permanent record of it has, for us, been well filled by the technique of staining infected cell cultures, learned under Dr. J. F. Enders and Dr. T. H. Weller in Boston.

Being solely responsible, over many years, for the validity of poliovirus vaccine safety tests we could at any time demonstrate the evidence on which our judgments were based, and we were able to reveal the presence of some prototype viruses that would otherwise have gone undetected. Furthermore, the technique has been invaluable in our investigations of outbreaks of viral diseases, and in the case of Marburg virus it opened up a way of studying a dangerous agent whose presence in cell cultures was being missed.

It has been a source of constant surprise to us that the staining of infected cell cultures is not more widely practiced. To some it is a simple and time-saving procedure; to others it appears to be complicated and time-consuming. To us its advantages greatly outweigh its limitations. The suggestion has been made on a number of occasions that we should produce an illustrated manual for the guidance of those wishing to use stained cultures for virus studies, and this presentation is in response to such requests. We have tried, however, to go beyond a bare manual by including some information that may not be otherwise readily available to the reader. In order to provide sufficient illustrations to indicate the wide range of cytopathic changes, it has been necessary to limit the size of the pictures to the minimum, but we trust that they will serve to convince the reader that the technique can be of great practical value for the recognition of many different viruses.

Hubert H. Malherbe
Margaret Strickland-Cholmley

THE AUTHORS

Hubert Malherbe received his M.A. degree in English Literature at Cambridge University, and qualified in medicine at the University of the Witwatersrand in South Africa, holding the degrees of M.B., B.Ch., and M.D.. During the Second World War he served with the South African Medical Corps in East Africa, and subsequently in the Viral and Rickettsial Laboratories of the South African Institute for Medical Research producing vaccines for the Allied Forces. In 1953 Dr. Malherbe was granted a Rockefeller Foundation Fellowship to study virological cell culture techniques under Drs. J. F. Enders and T. H. Weller in Boston. For the following 20 years he was in charge of the safety control of poliovirus vaccine produced in South Africa, which involved the detection of a wide variety of viruses of the African vervet monkey. In addition he carried out numerous studies on other viruses including the Marburg agent. Since 1974 Dr. Malherbe has been a resident in the U.S., and is in the Department of Microbiology at the University of Texas Health Science Center at San Antonio, where he teaches and conducts research on viruses, mycoplasmas, and chlamydiae.

Margaret Strickland-Cholmley was educated in South Africa and holds a diploma in virology. Since 1960 she has worked with Dr. Malherbe on vaccine control and on human and animal virus research, including studies on the Marburg virus and on prototype viruses isolated from vervet monkeys. She is also on the faculty of the Department of Microbiology at the University of Texas Health Science Center at San Antonio, teaching in the medical and dental schools as well as collaborating with Dr. Malherbe in research on viruses, mycoplasmas, and chlamydiae.

TABLE OF CONTENTS

Chapter 1

INTRODUCTION

I. CYTOPATHOLOGY AND CYTOPATHIC CHANGES

The propagation of viruses in vitro is now a major procedure in diagnosis and research, and examination of infected cell cultures constitutes an important part of cytopathology. It is the aim of this monograph to assist the virologist at the bench with identification of agents in cell cultures, particularly where it is essential to know in which main group of viruses an isolate belongs. It will also help in the detection of the presence of more than one virus in a cell culture, revealing early what could later become a source of considerable confusion and unnecessary work.

The term "cytopathic effect," or CPE, is often applied only to gross overt cellular changes in unstained cultures seen with the naked eye or with the low power of the light microscope, yet correctly it should comprise all degrees of cell damage, some of which are demonstrable only by means of specialized techniques. In this monograph we will consider mainly changes observable by light microscopy in cultures stained by methods that are well within the capability of any virological laboratory. It is our purpose to illustrate the range of information that can be obtained in this way, and to open a window or two, perhaps, for those who hold that, "Generally there is little advantage in staining infected cell cultures in an attempt to define the virus."[1]

Cytopathogenic and cytopathic are words that have frequently been applied indiscriminately both to agents causing cell damage and to the damage produced by them, although strictly speaking the terms are not synonymous. It has been proposed[2] that only "cytopathic" be used, and we have followed this practice.

Advances in sophisticated techniques at the subcellular and molecular levels have resulted in a prodigious outpouring of information about the behavior of viruses in cell cultures. From it are emerging important insights into the pathogenesis of disease, but less complex criteria by which practical decisions must daily be made in the laboratory tend to be neglected. We shall therefore try to convey the extent to which stained cultures can be used to expedite and control virological investigations, without attempting to present detailed interpretations of the cytochemical changes involved.

During the past three decades there have appeared a number of review articles on viral pathogenicity and cytopathology in cell cultures. A few of these are listed in the bibliography and can be consulted for further information.[3-6] In addition, the monograph by Cheville[7] provides a comprehensive description of the pathogenesis of virus infections in man and animals, with illustrations of histological and electron microscopical sections.

II. ARRANGEMENT OF THE TEXT

Our concern is primarily with practical aspects of the subject, and the material is accordingly presented with these in mind. After discussion of the general nature of cytopathic effects and of the preparation of stained cultures, we will present, with the aid of color photomicrographs, those changes observed mainly in the cytoplasm, followed by those occurring chiefly in the nucleus, and then those involving both cytoplasm and nucleus. Finally, we will consider multiple virus infections and nonviral contaminants of cell cultures.

III. THE ILLUSTRATIONS

Most of the stained cultures shown in the illustrations were prepared in the authors' laboratories at the Poliomyelitis Research Foundation, Johannesburg, South Africa, the Southwest Foundation for Research & Education, San Antonio, Texas, and the University of Texas Health Science Center at San Antonio.

The majority of the photomicrographs were taken by us with an Asahi Pentax® ME single-lens reflex camera in combination with a research model Zeiss®, microscope, using 35 mm Kodachrome® 25 film. We are indebted to Mr. Clyde Dexter of Dexter Instrument Company, San Antonio, for the loan of a Zeiss® Plan 100× objective.

We wish also to make the following acknowledgments: Dr. Sonja M. Buckley, Yale Arbovirus Research Unit, for the stained culture of Lassa virus shown in Figure 41;* Dr. Betty Edmond, Department of Pediatrics, University of Texas Health Science Center at San Antonio, for the culture of human embryo lung cells infected with varicella virus which we stained for Figure 78; Dr. J. H. S. Gear, South African Institute for Medical Research, for the conjunctival scraping infected with *Chlamydia trachomatis* shown in Figure 10; Dr. Donald E. Kahn, Biological Research Division, Pitman-Moore Inc., for sending us the CRFK feline kidney cells and the *feline panleukopenia* virus which we inoculated into them for Figure 49; Dr. Stewart McConnell, College of Veterinary Medicine, Texas A & M University, for the cultures of bluetongue and monkeypox viruses shown in Figures 32 and 38; Dr. Kendall O. Smith, Department of Microbiology, University of Texas Health Science Center at San Antonio, for the photomicrographs of cultures prepared and stained by him with fluorescein and with acridine orange, shown in Figures 68 and 76. The magnification given for each figure is that of the final reproduction.

REFERENCES

1. **Herrmann, E. C.**, New concepts and developments in applied diagnostic virology, *Prog. Med. Virol.*, 17, 221, 1974.
2. **Tyrrell, D. A., Stuart-Harris, C. H., and Andrewes, C. H**, Cytopathogenic vs. cytopathic, *Virology* 8, 539, 1959.
3. **Enders, J. F.**, Cytopathology of virus infections (particular reference to tissue culture studies), *Annu. Rev. Microbiol.*, 8, 473, 1954.
4. **Bablanian, R.**, Mechanisms of virus cytopathic effects, in *Microbial Pathogenicity in Man and Animals*, Cambridge University Press, London, 1972, 359.
5. **Smith, H.**, Mechanisms of virus pathogenicity, *Bacteriol. Rev.*, 36, 291, 1972.
6. **Bablanian, R.**,Structural and functional alterations in cultured cells infected with cytocidal viruses, *Prog. Med. Virol.*, 19, 40, 1975.
7. **Cheville, N. F.**, Cytopathology in viral diseases, *Monogr. Virol.*, 10, 1975.

* All figures appear following page 58.

Chapter 2

GENERAL CONSIDERATIONS OF CYTOPATHIC CHANGES

I. NONVIRAL TOXIC CHANGES

Cultures occasionally contain areas in which the cells show nuclear pyknosis or chromatin clumping, with degeneration of the cytoplasm (Figure 5).* These changes may be due to bacteria that are incompletely suppressed by antibiotics in the nutrient medium, and they are sufficiently distinctive to be differentiated from virus-induced injury.

Nuclear changes closely resembling those produced by viruses may be present in cells adjacent to contaminants such as amoebae, and they probably represent the effects of a toxic substance (Figure 6).

II. VIRAL TOXIC CHANGES

Cells can be injured by viral components that are sometimes referred to as viral toxins, of which the following are examples. The penton base antigen of adenoviruses causes cell rounding and detachment;[1-4] cell death may be produced by a reovirus outer capsid protein which retains its activity after the virus has been rendered noninfectious by ultraviolet irradiation;[5,6] paramyxoviruses possess hemolysins associated with the viral envelope,[7,8] as well as a cell-fusing factor that remains active after viral infectivity has been abolished by ultraviolet irradiation or treatment with betapropiolactone.[9,10] The fusing factor derived from the parainfluenza Sendai virus has been widely used for the production of heterokaryons since the introduction of this technique by Harris and Watkins.[11]

These subvirion parts probably contribute towards the entire cytopathic process in each virus infection. During abortive cycles in which little or no complete virus is formed, viral components accumulate in the cell and may be responsible for the damage observed.

Similar changes probably occur in the course of in vivo infections, but some tissue injury also appears to result from the deposition of viral antigen-antibody complexes and the activation of complement.[12,13]

III. INFECTION BY "NONCYTOPATHIC" VIRUSES

Viruses may infect cells without causing overt damage. Replication and discharge of virions occur in the absence of gross visible signs in the culture, and additional tests are necessary to reveal the presence of virus. These include hemagglutination with the supernatant nutrient fluid, hemadsorption by infected cells, interference with challenge by a known cytopathic virus, immunofluorescence or enzyme-linked immunosorbent assay to detect viral antigens, staining with acridine orange dye to locate and identify nucleic acids, and electron microscopy to render visible virus particles and subcellular changes not seen by light microscopy.

In other instances persistent infection involving only a few cells in the culture results in cytopathic change so limited that the cell sheet appears to be undamaged. More subtly, virus may remain within cells in a latent or cryptic form, and replication will

* All figures appear following page 58.

require the addition of chemical inducers, superinfection by a helper virus, or incorporation of the infected cells into a heterokaryon in order to provide metabolic systems lacking in the originally infected cells.

IV. INFECTION BY CYTOPATHIC VIRUSES

Cellular changes resulting from virus infection can be observed in several ways. Healthy cells cultured in vitro lower the pH of the nutrient fluid, and this alteration is made visible by incorporating the indicator dye phenol red in the medium. When virus infection leads to rapid destruction of cells, failure of the medium to change color reflects this activity. It should be noted, however, that slow intracellular changes can occur without noticeably affecting the pH of the medium.

Gross morphological changes may be visible under the low power of the light microscope as swelling of individual cells, as rounding with ultimate detachment from the substrate, or as giant cell formation. Areas of such changes may be visible to the naked eye, and when infected monolayers are covered with an overlay of agar or similar substance, such areas are seen as circumscribed plaques. Acting on the analogy of plaques caused by bacteriophage in cultures of susceptible bacteria, Dulbecco[14] in 1952 produced for the first time visible plaques in monolayer cell cultures infected with an animal virus under an agar overlay. This technique is now extensively used for the macroscopic detection of viral mutants, for virus assay, and for purification by selection of the progeny of single virus particles.

Morphological changes without cell destruction can occur when a virus with oncogenic potential infects susceptible cells. This transformation may be accompanied by the production of new antigens and the capacity to proliferate as tumor cells when inoculated into animals of the homologous species. The transformation of hamster cells in vitro by the simian adenovirus SA 7 is illustrated in Figure 65.

The formation of virus-induced multinucleated giant cells does not usually involve nuclear division. In some infections the cytoplasmic membranes between cells in contact with each other are dissolved, and fusion results. In such cases the nuclei tend to drift centrally, and the giant cell is called a syncytium (Figure 44). In other infections, ingestion of one cell by another through the process of cytophagia or cytophagocytosis produces a multinucleated cell, but dissolution of the cytoplasmic membrane of each ingested cell does not occur until phagocytosis has been completed (Figure 64). Cytopathic changes manifested as intracellular inclusions will be discussed in the following section.

V. INCLUSIONS

In his perspicacious article on the problem of intranuclear inclusions in virus diseases, Cowdry[15] drew attention to the fact that cytologists and virologists use the word "inclusion" differently. The former apply it to all intracellular structures other than normal organelles, comprising also substances like secretory granules and pigments that have nothing to do with viruses. Virologists, as Cowdry noted, have simply appropriated the term for use in a narrower sense. For the purpose of this study we shall regard inclusions as cytoplasmic or nuclear accumulations of proteins; lipids; carbohydrates; nucleic acids or their components that are produced as the result of virus infection and the consequent metabolic disturbance in the cell, or are produced through the action of nonviral factors leading to similar metabolic dysfunction. In the latter category are the chromophilic nuclear bodies produced by metabolic inhibitors such as actinomycin D which selectively inhibits the synthesis of cellular RNA (Figures 7 and 8), or by toxins (Figure 6). Criteria of shape, texture, and staining affinity are used to distinguish virus-induced inclusions from other structures such as the nuclear vacuoles seen in some continuously cultivated line cells (Figure 9), or the ingested debris commonly observed in the cytoplasm of cultured cells.

Most viruses, being less than 200 nm in diameter, are below the resolving power of the light microscope. Virus-induced inclusions are therefore not collections of visible virions, but consist of metabolic products in which virus particles may or may not be present. In contrast, the cytoplasmic inclusion seen in the chlamydia-infected cell of trachoma comprises visible elementary bodies embedded in a glycogen matrix and is, indeed, a colony of the infectious agent (Figure 10).

Cowdry divided nuclear inclusions into two types. Type A inclusions are usually acidophilic and occupy the greater part of the nucleus. The chromatin is marginated, and the nuclear reaction is total and proceeds to complete degeneration. Type B inclusions are localized acidophilic bodies embedded in a relatively unaltered nucleoplasm; the chromatin is not marginated, and the nucleus seldom undergoes complete degeneration. Cowdry was of the opinion that a viral etiology was probable for most Type A inclusions but less likely for those of Type B. He noted that both types could be present in the same specimen, but regarded this as exceptional. Cowdry was concerned entirely with histological preparations, without benefit of the detailed examination of cells afforded by cell cultures. Today it is a simple matter to demonstrate that small and large nuclear inclusions, with or without chromatin margination, frequently occur in the same infected culture (Figure 11) and that there is a much wider range of morphological and tinctorial properties displayed by nuclear inclusions than Cowdry observed. It seems appropriate now not to divide nuclear inclusions simply into Types A and B. The distinctive appearances of different virus inclusions will be discussed later in the relevant chapters of the text.

REFERENCES

1. **Pereira, H. G. and Kelly, B.**, Dose-response curves of toxic and infective actions of adenovirus in HeLa cell cultures, *J. Gen. Microbiol.*, 17, 517, 1957.
2. **Pereira, H. G.**, A protein factor responsible for the early cytopathic effect of adenoviruses, *Virology*, 6, 601, 1958.
3. **Everett, S. F. and Ginsberg, H. S.**, A toxinlike material separable from Type 5 adenovirus particles, *Virology*, 6, 770, 1958.
4. **Davis, B. D., Dulbecco, R., Eisen, H. N., Ginsberg, H. S., Wood, W. B., and McCarty, M.**, Adenoviruses, in *Microbiology*, 2nd ed., Harper & Row, Hagerstown, Md., 1973, 1226.
5. **Loh, P. C. and Oie, H. K.**, Growth characteristics of reovirus type 2: ultraviolet light inactivated virion preparations and cell death, *Arch. Gesamte Virusforsch.*, 26, 197, 1969.
6. **Subasinghe, H. A. and Loh, P. C.**, Reovirus cytotoxicity: some properties of the UV-irradiated reovirus and its capsid proteins, *Arch. Gesamte Virusforsch.*, 39, 172, 1972.
7. **Morgan, H. R., Enders, J. F., and Wagley, P. F.**, A hemolysin associated with the mumps virus, *J. Exp. Med.*, 88, 503, 1948.
8. **Jawetz, E., Melnick, J. L., and Adelberg, E. A.**, Adenovirus family, in *Review of Medical Microbiology*, 13th ed., Lange, Los Altos, Calif., 1978, 440.
9. **Neff, J. M. and Enders, J. F.**, Further observations on replication of type 1 poliovirus in naturally resistant fused cell cultures, *Perspect. Virol.*, 6, 39, 1968.
10. **Pedreira, F. A. and Tauraso, N. M.**, Effect of beta-propiolactone, ultraviolet irradiation, and formalin on Sendai virus cell-fusion factor, *Arch. Gesamte Virusforsch.*, 28, 361, 1969.

11. **Harris, H. and Watkins, J. F.**, Hybrid cells derived from mouse and man: artificial heterokaryons of mammalian cells from different species, *Nature (London)*, 205, 640, 1965.
12. **Oldstone, M. B. and Dixon, F. J.**, Immune complex disease in chronic viral infections, *J. Exp. Med.*, 134, 32s, 1971.
13. **Fenner, F., McAuslan, B. R., Mims, C. A., Sambrook, J., and White, D. O.**, *The Biology of Animal Viruses*, 2nd ed., Academic Press, New York, 1974, 346.
14. **Dulbecco, R.**, Production of plaques in monolayer tissue cultures by single particles of an animal virus, *Proc. Natl. Acad. Sci. U.S.A.*, 38, 747, 1952.
15. **Cowdry, E. V.**, The problem of intranuclear inclusions in virus diseases, *Arch. Pathol.*, 18, 527, 1934.

Chapter 3

PREPARATION OF CELL CULTURES FOR THE STUDY OF CYTOPATHIC CHANGES

I. SELECTION AND IDENTIFICATION OF CELLS

The cytopathic effect produced by a virus can vary to some extent with the type of cell used. The changes we shall illustrate have elements that are common to different kinds of cells, but it is important for the virologist to develop a thorough acquaintance with the variations in cytopathology that occur in the particular range of cells he uses.

No single type of cell supports the growth of all viruses, and therefore each laboratory must select those cells best suited to its special needs. The literature is replete with papers describing the uses of various cells for the cultivation of viruses. We will not attempt an extensive review of the subject, but will indicate a few studies to which references can be made for special requirements.[1-10] Methods for the preparation of cultures, the particular applications of different cell types, and references to the origins of a number of widely used cell lines are to be found in the comprehensive presentation by Schmidt.[11]

The suitability of any cell type for virus cultivation depends on its capacity to support all stages in the infective and replicative cycles. These include the attachment of virus particles to specific receptors on the surface of the cell, penetration, uncoating, synthesis of nucleic acids and proteins, and final discharge of virus with, in some instances, an envelope derived from the cytoplasmic membrane. Failure of the cell to participate successfully in any one of these stages will result in either complete or partial resistance. In the latter case severe damage to the cell can occur, and the cytopathic changes may be quite characteristic of the particular infection, as in the incomplete replication of human adenoviruses in simian cells.

Holland[12] investigated the development in cultured cells of surface receptors for enteroviruses that were not present in vivo, and concluded that synthesis of receptors may be suppressed normally in various cells in the body but becomes derepressed when the cells are cultured in vitro. It was shown by Evans et al.[13] that whereas poliovirus multiplies readily in monkey testis cells cultivated in vitro, little or no viral growth occurs in vivo following intratesticular inoculation of the virus. Comparable results were obtained by Kaplan[14] with regard to the renal cells of monkeys whose kidneys had been inoculated in vivo with poliovirus. Holland[12] also found that human amnion, cultured as such without separation of the epithelial cells from the supporting collagen membrane, was refractory to poliovirus; but trypsin-dispersed epithelial cells from the same amnion, grown as a monolayer on glass, became highly susceptible to the virus. Frothingham[15] noted that monolayers of human amniotic cells cultivated for less than 30 days were markedly resistant to infection with Sindbis virus and poliovirus, in comparison with the sensitivity of amnion cells cultured for longer periods prior to challenge with virus.

The affinity or tropism of viruses for certain cells in the body, as manifested in hepatitis, encephalitis, orchitis, etc. indicates that specific surface receptors do exist in vivo, and the particular predilection shown by coxsackie group B viruses for neonatal organs may be related to the relative abundance of receptors at that age.[16]

Thus cytopathic changes can be significantly influenced by the variable presence or absence of surface receptors, depending in part on the age of the culture. In addition, the extent to which each virus is capable of inducing interferon will modify the infectious process in the cell culture.

Three kinds of cell cultures are commonly used for the isolation of viruses, namely, primary cultures, cell strains, and cell lines. Hayflick and Moorhead[17] proposed that the term "strain" be used for a population of cells subcultured more than once, in which the chromosomal karyotype characterizing the tissue of origin, that is, the normal diploid complement of chromosomes, is preserved. They suggested that the term "line" be used for a cell population that has been serially subcultivated in vitro for indefinite periods of time and shows a departure from the normal chromosome number.

A primary culture is a heterogeneous collection of cells drawn from the various components of a complex tissue. The cells in it are all initially euploid and diploid, that is, they all contain the normal number of chromosomes made up of the combined half or haploid contributions from sperm and ovum which together comprise the double or diploid number. Upon repeated subculture a change takes place, and the cells may then undergo a crisis which ends in complete degeneration, or they may become capable of indefinite cultivation as a line which is characterized by variations in the chromosome number. Such heteroploidy is marked by morphological changes in the affected nuclei, and it may also be accompanied by a reduction in the sensitivity of the culture to certain virus infections. Similar cell variations can arise even in cell populations cloned as the progeny of a single cell.

Thus it comes about that some subcultivations of cell lines may differ in virus sensitivity from the parent population. For example, the HeLa cell clone S3 was noted by Vogt and Dulbecco to have become resistant to poliovirus.[18] In this instance the resistant cells had a fusiform shape and may have represented the selection of a type of cell that had been present in the line for some time. Hayflick and Moorhead[17] observed that of 25 human fetal cell strains successfully cultured by them, all, without exception, were of the fibroblast cell type from about the fifth subcultivation, including those from the kidney. It is common practice to refer to elongated cells as fibroblasts rather than fibroblast-like, and definitive tests to establish the true identity of cultured cells are very seldom done. It may be as well to bear in mind the admonition of Farnes[19] that all that spindles is not a fibroblast.

Alteration in the behavior of a cell population may also be due to cross-contamination between cell cultures. This is notoriously common when specific precautions against it are neglected. Experimental results become vitiated through erroneous use of cells having the right label but the wrong ancestry. A number of studies have been undertaken to find distinctive markers for various cell types and by means of them to detect contamination or mislabeling of cultures.[19-25] Reference to these will readily demonstrate the ease with which cell lines have been mixed up.

II. COVERSLIP CULTURES AND ROLLER TUBES

When Enders et al.[26, 27] first reported in 1949 that poliovirus multiplied and caused morphological changes in nonneural cells, their cultures comprised minced tissue suspended in nutrient fluid in flasks from which fragments were removed for histological fixation, sectioning, and staining. They then proceeded[28] to utilize roller tubes, on the walls of which were placed tissue fragments embedded in chick plasma clots according to the method used by Feller et al.[29] in 1940 for the propagation of vaccinia virus. The roller tube, originally devised by Carrel in 1913,[30] was independently developed in the United States by Gey[31] in 1933. Cultivation of the chlamydia of lymphogranuloma venereum in roller tubes was achieved by Gey and Bang[32] in 1939, but the first use of the technique for propagating a true virus was by Feller et al.[29]

In 1953 Weller[33] cultured tissue explants on coverslips coated with chicken plasma that were placed free in roller tubes for the first successful isolation of varicella-zoster virus. The coverslip cultures were then stained with hematoxylin and eosin to demonstrate the nuclear inclusions. Measles virus was first propagated by Enders and Peebles[34] who used monolayer roller tube cultures of human and monkey kidney cells prepared according to the method described by Youngner.[35] They also employed a technique developed by Cheatham in which the layer of cells was fixed with formalin, covered with a collodion membrane, and then detached from the tube for staining and mounting on a slide.

The dispersion of cells by mechanical means, and their cultivation on a glass surface, had been developed by Earle and his colleagues at the National Cancer Institute.[36, 37] Dulbecco and Vogt[38] used trypsin to prepare monkey kidney cell monolayers for the propagation of poliovirus, and Youngner extended the technique in order to produce larger volumes of standardized cell suspensions cultured in tubes or bottles which could be incubated in an atmosphere of 3% carbon dioxide.[35] The ease with which trypsin-dispersed cells adhered to glass surfaces led subsequently to the use of coverslips seeded with cell suspensions, either in petri dishes incubated under partial tension of carbon dioxide or lying free within closed tubes.

In 1954 Leighton[39] modified the roller tube by flattening part of one side in order to support a cellulose sponge matrix for the cultivation of tumor cells. Inside the same tube he placed a "flying coverslip" culture which served to compare growth of the cells in the thin coverslip culture with that of the cells in the sponge matrix. The purpose of the flattened area was to provide an uncurved surface for better examination of the cells growing in the matrix, not to support the coverslip which moved freely in the roller tube. The Leighton tube, however, soon came into use for holding monolayer coverslip cultures incubated in a stationary position, although there is no real need for a flattened surface to support the coverslip.

It has been the experience of competent workers that the roller tube can significantly advance both viral growth and cytopathic effects.[3, 11, 40-42] In addition, the movement reduces the amount of debris that settles on the culture and is ingested by the cells, resulting in a clearer evaluation of cytopathology than is afforded by the stationary culture. The cells on the coverslip can be readily examined for overt signs of change while still in the tube; and since they do not come into contact with the curved wall of the tube, they remain undamaged by the rotation.

REFERENCES

1. **Enders, J. F.,** Cytopathology of virus infections (Particular reference to tissue culture studies), *Annu. Rev. Microbiol.*, 8, 473, 1954.
2. **Takemoto, K. K. and Lerner, A. M.,** Human amnion cell cultures; susceptibility to viruses and use in primary virus isolations, *Proc. Soc. Exp. Biol. Med.*, 94, 179, 1957.
3. **Tyrrell, D. A. and Parsons, R.,** Some virus isolations from common colds. III. Cytopathic effects in tissue cultures, *Lancet*, 1, 239, 1960.
4. **Hamre, D.,** Critique of methods for recovery of respiratory viruses, *Am. Rev. Respir. Dis.*, 88, 306, 1963.
5. **Votava, M. and Tyrrell, D. A.,** The possible value of certain cells for the propagation of respiratory viruses, *Arch. Gesamte Virusforsch.*, 29, 253, 1970.
6. **Cooney, M. K.,** Relative efficiency of cell cultures for detection of viruses, *Health Lab. Sci.*, 10, 294, 1973.
7. **McSwiggan, D. A., Darougar, S., Rahman, A. F., and Gibson, J. A.,** Comparison of the sensitivity of human embryo kidney cells, HeLa cells, and WI 38 cells for the primary isolation of viruses from the eye, *J. Clin. Pathol.*, 28, 410, 1975.

8. **Schmidt, N. J., Ho, H. H., Riggs, J. L., and Lennette, E. H.,** Comparative sensitivity of various cell culture systems for isolation of viruses from wastewater and fecal samples, *Appl. Environ. Microbiol.*, 36, 480, 1978.
9. **Friedman, H. M. and Koropchak, C.,** Comparison of WI-38, MRC-5, and IMR-90 cell strains for isolation of viruses from clinical specimens, *J. Clin. Microbiol.*, 7, 368, 1978.
10. **Meguro, H., Bryant, J. D., Torrence, A. E., and Wright, P. F.,** Canine kidney cell line for isolation of respiratory viruses, *J. Clin. Microbiol.*, 9, 175, 1979.
11. **Schmidt, N. J.,** Tissue culture technics for diagnostic virology, in *Diagnostic Procedures for Viral and Rickettsial Infections,* 4th ed., Lennette, E. H. and Schmidt, N. J., Eds., American Public Health Assoc., New York, 1969, chap. 3.
12. **Holland, J. J.,** Receptor affinities as major determinants of enterovirus tissue tropisms in humans, *Virology*, 15, 312, 1961.
13. **Evans, C. A., Byatt, P. H., Chambers, V. C., and Smith, W. M.,** Growth of neurotropic viruses in extraneural tissues. VI. Absence of in vivo multiplication of poliomyelitis virus, types 1 and 2, after intratesticular inoculation of monkeys and other laboratory animals, *J. Immunol.*, 72, 348, 1954.
14. **Kaplan, A. S.,** Comparison of susceptible and resistant cells to infection with poliomyelitis virus, *Ann. N. Y. Acad. Sci.*, 61, 830, 1955.
15. **Frothingham, T. E.,** Effect of aging and rotation on human amnion cell response to polio and Sindbis viruses, *Proc. Soc. Exp. Biol. Med.*, 100, 505, 1959.
16. **Kunin, C. M.,** Virus-tissue union and the pathogenesis of enterovirus infections, *J. Immunol.*, 88, 556, 1962.
17. **Hayflick, L. and Moorhead, P. S.,** The serial cultivation of human diploid cell strains, *Exp. Cell. Res.*, 25, 585, 1961.
18. **Vogt, M. and Dulbecco, R.,** Properties of a HeLa cell culture with increased resistance to poliomyelitis virus, *Virology*, 5, 425, 1958.
19. **Farnes, P.,** Histochemical approaches to cell characterization in vitro, *Natl. Cancer Inst. Monogr.*, 26, 199, 1967.
20. **Stulberg, C. S.,** Extrinsic cell contamination of tissue cultures, in *Contamination in Tissue Culture,* Fogh, J., Ed., Academic Press, New York, 1973, chap. 1.
21. **Hsu, S. H., Schacter, B. Z., Delaney, N. L., Miller, T. B., McKusick, V. A., Kennett, R. H., Bodmer, J. G., Young, D., and Bodmer, W. F.** Genetic characteristics of the HeLa cell, *Science,* 191, 392, 1976.
22. **Nelson-Rees, W. A. and Flandermeyer, R. R.,** HeLa cultures defined, *Science*, 191, 96, 1976.
23. **Povey, S., Hopkinson, D. A., Harris, H., and Franks, L. M.,** Characterisation of human cell lines and differentiation from HeLa by enzyme typing, *Nature (London)*, 264, 60, 1976.
24. **Nelson-Rees, W. A. and Flandermeyer, R. R.,** Inter- and intraspecies contamination of human breast tumor cell lines HBC and BrCa5 and other cell cultures, *Science*, 195, 1343, 1977.
25. **O'Brien, S. J., Kleiner, G., Olson, R., and Shannon, J. E.,** Enzyme polymorphisms as genetic signatures in human cell cultures, *Science*, 195, 1345, 1977.
26. **Enders, J. F., Weller, T. H., and Robbins, F. C.,** Cultivation of the Lansing strain of poliomyelitis virus in cultures of various human embryonic tissues, *Science*, 109, 85, 1949.
27. **Weller, T. H., Robbins, F. C., and Enders, J. F.,** Cultivation of poliomyelitis virus in cultures of human foreskin and embryonic tissues, *Proc. Soc. Exp. Biol. Med.*, 72, 153, 1949.
28. **Robbins, F. C., Enders, J. F. , and Weller, T. H.,** Cytopathogenic effect of poliomyelitis viruses in vitro on human embryonic tissues, *Proc. Soc. Exp. Biol. Med.*, 75, 370, 1950.
29. **Feller, A. E., Enders, J. F., and Weller, T. H.,** The prolonged coexistence of vaccinia virus in high titre and living cells in roller tube cultures of chick embryonic tissues, *J. Exp. Med.*, 72, 367, 1940.
30. **Carrel, A.,** *Berl. Klin. Wochenschr.*, 50, 1097, 1913; as cited by **Gey, G. O.,** *Am. J. Cancer*, 17, 752, 1933.
31. **Gey, G. O.,** An improved technic for massive tissue culture, *Am. J. Cancer*, 17, 752, 1933.
32. **Gey, G. O. and Bang, F. B.,** Experimental studies on the cultural behavior and the infectivity of Lymphopathia venerea virus maintained in tissue culture, *Bull. Johns Hopkins Hosp.*, 65, 393, 1939.
33. **Weller, T. H.,** Serial propagation in vitro of agents producing inclusion bodies derived from varicella and herpes zoster, *Proc. Soc. Exp. Biol. Med.*, 83, 340, 1953.
34. **Enders, J. F. and Peebles, T. C.,** Propagation in tissue cultures of cytopathogenic agents from patients with measles, *Proc. Soc. Exp. Biol. Med.*, 86, 277, 1954.
35. **Youngner, J. S.,** Monolayer tissue cultures. I. Preparation and standardization of suspensions of trypsin-dispersed monkey kidney cells, *Proc. Soc. Exp. Biol. Med.*, 85, 202, 1954.
36. **Evans, V. J., Earle, W. R., Sanford, K. K., Shannon, J. E., and Waltz, H. K.,** The preparation and handling of replicate tissue cultures for quantitative studies, *J. Natl. Cancer Inst.*, 11, 907, 1951.
37. **Shannon, J. E. and Earle, W. R.,** Qualitative comparison of the growth of chick-heart and strain L fibroblasts planted as suspensions on pyrex glass and perforated cellophane substrates, *J. Natl. Cancer Inst.*, 12, 155, 1951.

38. **Dulbecco, R. and Vogt, M.,** Plaque formation and isolation of pure lines with poliomyelitis viruses, *J. Exp. Med.*, 99, 167, 1954.
39. **Leighton, J.,** The growth patterns of some transplantable animal tumors in sponge matrix tissue culture, *J. Natl. Cancer Inst.*, 15, 275, 1954.
40. **Robbins, F. C., Weller, T. H., and Enders, J. F.,** Studies on the cultivation of poliomyelitis viruses in tissue culture. II. The propagation of the poliomyelitis viruses in roller-tube cultures of various human tissues, *J. Immunol.*, 69, 673, 1952.
41. **Lerner, A. M., Cherry, J. D., and Finland, M.,** Enhancement of cytopathic effects of reoviruses in rolled cultures of rhesus kidney, *Proc. Soc. Exp. Biol. Med.*, 110, 727, 1962.
42. **Canchola, J., Vargosko, A. J., Kim, H. W., Parrott, R. H., Christmas, E., Jeffries, B., and Chanock, R. M.,** Antigenic variation among newly isolated strains of parainfluenza Type 4 virus, *Am. J. Hyg.*, 79, 357, 1964.

Chapter 4

FIXATIVES AND STAINS

I. INTRODUCTION

The procedure of fixing and staining monolayer coverslip cultures was a direct extension of the histological processing of cultured tissue fragments carried out by Enders and his colleagues[1] when they first observed that changes occurred in nonneural cells infected with poliovirus. It is therefore not surprising that use of the same reagents has continued with few additions. The fixatives most widely employed are Zenker's and Bouin's fluids, and the staining methods commonly used are hematoxylin-eosin and May-Grunwald-Giemsa.

II. FIXATIVES

Cowdry[2, 3] had observed that the kind of fixative used in histological preparations influenced both the morphology and the staining reaction of nuclear inclusions, and he noted that the width of the halo around an inclusion depended on whether the fixative had led to swelling or shrinkage of the nucleus. In their study of the cellular changes produced by the simian herpes B virus, Reissig and Melnick[4] confirmed Cowdry's comment about shrinkage, and they followed fixation with Zenker's fluid by immersion in 80% alcohol to increase retraction of the inclusion material away from the marginated chromatin. This gave a more distinct appearance to the inclusions, and on the same basis we have routinely used overnight treatment with 70% alcohol in our fixation procedure.

Zenker's fluid contains mercuric chloride and potassium dichromate as protein coagulants, while Bouin's fluid has picric acid as the coagulant, with formaldehyde added as a noncoagulating fixative. Both fluids contain glacial acetic acid which is intended to counteract some of the shrinkage caused by the other constituents. This may seem anomalous in view of the advantages of shrinkage just mentioned, but not all the processes involved in fixation are fully understood, and the use of glacial acetic acid on empirical grounds may well be justified by the results.

Cowdry noted that fixation with osmic acid did not produce the coagulation of nuclear protein that results from fixation with Zenker's fluid. Reissig and Melnick also found that nuclear inclusion material was not clearly recognizable as such when osmium tetroxide alone was used to fix the cells. Thus while osmic acid remains one of the best fixatives for electron microscopy where it is essential to preserve relationships between cellular components, for light microscopy of infected cells it is not only unnecessary but also undesirable to avoid fixatives that coagulate and shrink proteins. Visual perception of structural changes depends largely upon contrasts of texture, shape, and tinctorial values; and fixatives that produce good contrasts are to be preferred for purely pragmatic reasons.

III. STAINS

All stains have selective cytochemical actions, and in order to detect the presence of lipids, proteins, carbohydates, and nucleic acids it is necessary to use a variety of stains which no single procedure employs. It is also important to use fixatives that preserve the substances one wishes to stain. Methods such as Lendrum's phloxine tartrazine combination have been devised specifically to accentuate virus inclusions, and although the effects can be vivid and distinctive, such techniques have specialized applications and are not versatile enough to detect a wide range of cytopathic changes.

May-Grunwald stain is equivalent to Jenner's stain,[5] and contains methylene blue eosinate. Giemsa's stain contains methylene blue, azure A and azure B, and eosin, these being present as the eosinates of azures A and B, and methylene blue eosinate. Giemsa is a neutral stain, which is defined as a compound of an acid dye and a basic dye, in which both the anion and the cation contain chromophore groups. May-Grunwald and Giemsa stains can be applied either in a two stage procedure,[6, 7] or as a single solution.[8] The tinctorial effects of these stains differ appreciably from those of the simpler hematoxylin-eosin procedure, but the additional information obtained may not warrant the extra care and expense. When used for cultures that must be dehydrated before mounting, the May-Grunwald or Giemsa techniques present some problems in removal of stain in the alcohol rinses, and the stained cultures may have to be air-dried and then rinsed with xylene before mounting.

Methylene blue and the azures are basic dyes, and structures that combine with them are basophilic. The observed color reactions, however, depend upon the nature of the fixative used and on the pH of the staining solution. Thus when Giemsa stain is applied it is necessary to know these conditions in order to interpret the color reactions correctly. When a structure stained with Giemsa is reported as basophilic, it is understood that it has stained blue under the particular, and usually unspecified, conditions of the experiment; but with another stain the structure may be shown to be acidophilic and stain red with eosin. In stating the reported staining reactions of inclusions we have usually used the terms employed by the authors of the papers, but these should be interpreted with caution. In some instances the same kind of inclusion has been described as basophilic or eosinophilic by different authors. The paper by Kato et al. (Reference 107 in Chapter 5) considerably clarified the position by showing that inclusions which were blue when fixed with methanol and stained with Giemsa, were bright red when fixed with Bouin's fluid and stained with hematoxylin-eosin. We have demonstrated that a similar situation exists with reovirus inclusions (Figures 23 and 24),* and also with rotavirus inclusions (Figures 26 and 29).

Hematoxylin itself does not stain cells, and it must first be converted by oxidation to hematein. This can be done either by prolonged exposure of hematoxylin solution to light and air, or more rapidly by treatment with sodium iodate. Hematein is a weak yellow dye, but it forms a purple-red lake with aluminum sulfate, and this binds to the DNA of the cell nucleus, turning blue during the alkaline water rinses. We have found that most commercially available brands of hematoxylin are satisfactory for preparing Ehrlich's acid hematoxylin solution.

Eosin stains proteins that occur chiefly in the cytoplasm but also to a lesser extent in the nucleus. For good results it is necessary to use the alcohol-soluble form, the best being that supplied by the British Drug Houses Company as BDH Spirit Soluble Eosin C.I. 45386. Methods for the preparation of Bouin's fluid, Ehrlich's hematoxylin, and eosin solution are given in the Appendix.

A satisfactory mounting medium should dry hard, adhere firmly to glass, remain clear, and not cause fading of the stains. All of these properties are to be found in Eukitt mountant which is made by O. Kindler of West Germany. The mounting medium DPX (Distrene-Plasticizer-Xylene)[9] is also reliable and easy to prepare. Neither of these media causes the fading of stains that follows the use of Permount. The coverslip is mounted with the cell culture on the underneath surface, thus providing a permanent specimen that should remain unaltered for an indefinite time.

* All figures appear following page 58.

REFERENCES

1. **Enders, J. F., Weller, T. H., and Robbins, F. C.**, Cultivation of the Lansing strain of poliomyelitis virus in cultures of various human embryonic tissues, *Science*, 109, 85, 1949.
2. **Cowdry, E. V.**, A comparison of the intranuclear inclusions produced by the herpetic virus and by Virus III in rabbits, *Arch. Pathol.*, 10, 23, 1930.
3. **Cowdry, E. V.**, The problem of intranuclear inclusions in virus diseases, *Arch. Pathol.*, 18, 527, 1934.
4. **Reissig, M. and Melnick, J. L.**, The cellular changes produced in tissue cultures by Herpes B virus correlated with the concurrent multiplication of the virus, *J. Exp. Med.*, 101, 341, 1955.
5. **Lillie, R. D.**, *H. J. Conn's Biological Stains*, 9th ed., Williams & Wilkins, Baltimore, 1977.
6. Armed Forces Institute of Pathology, *Manual of Histologic Staining Methods*, 3rd ed., Luna, L. G., Ed., McGraw-Hill, New York, 1968, 121.
7. **Allen, R., Sims, R. A., and Sulkin, S. E.**, Studies with cultured brown adipose tissue. I. Persistence of rabies virus in bat brown fat, *Am. J. Hyg.*, 80, 11, 1964.
8. **Strumia, M. M.**, A rapid universal blood stain, *J. Lab. Clin. Med.*, 21, 930, 1936.
9. **Baker, J. R.**, *Cytological Technique*, 5th ed., John Wiley & Sons, New York, 1966.

Chapter 5

CYTOPATHIC CHANGES AFFECTING MAINLY THE CYTOPLASM

I. PICORNAVIRUSES

The family Picornaviridae comprises four genera, namely, *rhinovirus, enterovirus, cardiovirus, and aphthovirus.*[1] Of these, the first two are of importance in human disease.

Rhinoviruses tend to have particular cultural requirements relating to temperature of incubation and to composition of nutrient medium.[2] Some strains show preference for certain human or monkey cells;[3-6] others are noncytopathic and require an interference test to reveal their presence.[7,8] Further information can be found in the review monograph by Tyrrell.[9]

The enteroviruses, comprising polio-, echo-, and coxsackieviruses, grow well in cell cultures with the exception of some of the group A coxsackieviruses which either do not produce overt cytopathic changes or do not grow at all in cell cultures, and which require the inoculation of newborn mice for their detection.[10]

Cells normally refractory to poliovirus, such as chick embryo and hamster cells, can support replication when penetration of the virus is facilitated through alteration of the cytoplasmic membrane by means of the cell fusion factor of Sendai virus.[11-13]

It is generally agreed that the site of picornavirus replication is in the cytoplasm, where it is marked by the appearance of a large round paranuclear zone which displaces and compresses the nucleus (Figures 12 and 13).* This area is hyaline and usually stains lightly with eosin. It is surrounded by denser cytoplasm in which basophilic bodies may be found (Figure 14) as noted by several early investigators.[14-17] This zone is sometimes referred to as an inclusion body, which it undoubtedly is; but its relative lack of staining capacity makes the term less appropriate for it than for most other inclusions. Picornavirus-infected cells may also have long cytoplasmic processes (Figure 15). Eventually the cells round up, become detached, and lyse through rupture of the cytoplasmic membrane, for which a cytotoxic double-stranded RNA viral product may be responsible.[1]

In their historic study on the cultivation of poliovirus, Robbins et al.[18] observed that in sectioned fragments of infected tissue there was loss of typical staining properties together with nuclear pyknosis and disintegration of the cells. Barski and his colleagues in Paris then reported the sequence of cytoplasmic changes in poliovirus-infected cells as seen by both direct and phase contrast light microscopy.[19,20] The appearance of small eosinophilic nuclear inclusions as early as 4 hr after infection with poliovirus was noted by Beale and his co-workers[21] and by Reissig et al.,[15] but Mayor[22] failed to detect viral antigen in such inclusions. Reports have appeared in the literature[23-26] concerning the presence of virus-like particles in the nuclei of picornavirus-infected cells, but such structures have not been consistently observed, and the main replicative cycle appears to take place in the cytoplasm.[27,28] Cells enucleated with cytochalasin B and subsequently infected with poliovirus can maintain the whole cycle; but the nucleus may influence viral growth, because enucleated cells support replication less efficiently than do cells with nuclei.[29,30] Very early structural changes in the nuclei of poliovirus-infected cells have been noted,[31] but the virus inhibits synthesis of cellular RNA and protein, and the changes may in part be due to this. It has been suggested, however,

* All figures appear following page 58.

that most of the morphological changes caused by picornaviruses may be due to the direct action of viral components rather than to the inhibition of cellular metabolism.[32,33] The similarity between early ultrastructural nucleolar changes caused by coxsackievirus A9 and those produced by actinomycin D in patas monkey kidney cells was noted by Weiss and Meyer.[34] In both instances interference with cellular RNA metabolism may be manifested in the same way. Distinctive nucleolar dispersion is seen in cells infected with certain simian picornaviruses such as SV 44 (Figure 16). Thus while the main cytopathic effect is seen in the cytoplasm, less obvious changes occur in the nucleus as well.

Damage to the nucleus may be delayed long enough for a considerable amount of eosinophilic inclusion material to accumulate. In the case of attenuated poliovirus strains, for example, faintly eosinophilic substance may fill the nucleus before distortion occurs. We have noted[35] that numerous well-developed nuclear inclusions form in vervet monkey kidney cell cultures infected simultaneously with picornavirus and spumavirus (simian foamy virus), both agents having cytoplasmic replicative cycles. This effect is illustrated in Figures 17 and 18.

The above considerations serve to emphasize that when infected cell cultures are observed only with the naked eye or with the low power of the light microscope (Figure 19) little can be inferred except that something is happening, whereas staining reveals many distinctive changes.

II. RHABDOVIRUSES

The rhabdovirus group includes rabiesvirus (lyssavirus) and the arthropodborne vesicular stomatitis virus. Marburg virus has not yet been finally classified, but because it is morphologically closer to the rhabdoviruses than to any other group, we have included it here. Bablanian[36] has reviewed studies on the changes produced in cell cultures by vesicular stomatitis virus, and in this section we will consider only rabies and Marburg viruses.

A. Rabiesvirus (Lyssavirus)

The cytosplasmic inclusion recognized by Negri[37] in 1903 as pathognomonic of rabies infection has proved to be of great diagnostic importance, matched in value only relatively recently by the immunofluorescence test which extends the perception of cytopathic changes beyond the limits of light microscopy. By this test the presence of antigen is detected when Negri bodies are not produced, as in the case of certain virus strains infecting skunks and bats. There have been instances in which Negri bodies were present but the immunofluorescence test appeared to be negative,[38,39] possibly owing to antigenic difference between the infecting strain and that used to prepare the labeled antibody.

The Negri body (Figure 20) is an exclusively intracytoplasmic structure having a clearly defined round or oval shape. The main mass of it has an acidophilic staining reaction, but basophilic internal bodies should be seen in it in order to distinguish it from nonrabies inclusions such as those of distemper virus in dogs.

Negri regarded the inclusion as a protozoon which he considered to be the cause of rabies, and he noted the variable numbers and sizes of the internal structures. Goodpasture[40] accepted that by definition the Negri body must have internal structures, and he applied the name lyssa body to the smaller inclusions in rabies-infected cells that lacked inner structures but in other respects were similar to Negri bodies. The term lyssa body is still sometimes used, but in essence all rabies-induced inclusions are the same.[41] Nevertheless, presence of the basophilic internal structures is of importance in differentiating rabies from other virus infections.

The matrix of the inclusion appears to consist largely of nucleocapsid material containing ribonucleic acid; and mature virus particles are not invariably present in the Negri body.[42-45] By examining alternate thick and thin sections of the same inclusions in infected mouse brain cells by light and electron microscopy, Miyamoto and Matsumoto[46] noted that the inner basophilic structures of the Negri body contained virus particles together with some cellular endoplasmic reticulum.

The propagation of rabiesvirus in cell cultures has been reviewed by Wiktor and Clark.[41] These workers have found that nearly every type of mammalian or avian cell in primary or line culture is capable of supporting the growth of rabiesvirus. The BHK 21 line of neonatal Syrian hamster kidney cells in particular has given high yields of virus. Replication can occur in the absence of a gross cytopathic change in the cell culture. The presence of intracellular virus can be demonstrated by means of immunofluorescence, but cytoplasmic inclusions resembling Negri bodies have been observed in some cultures stained by Seller's or May-Grunwald-Giemsa technique.[41,47]

B. Marburg Virus

In 1967 a severe and apparently new hemorrhagic disease affected 25 laboratory workers in Germany and Yugoslavia, causing seven deaths. These persons had been involved in the handling of African vervet monkeys or the processing of their kidneys for cell cultures. In addition, there were six secondary cases. Four batches of monkeys sent from Uganda were implicated as the source of the infection.

Intensive investigations in laboratories in several countries led to the isolation of a filamentous and sometimes curved virus whose final classification has not yet been decided. The outbreak in Europe and the investigations arising from it have been well documented.[48,49]

The monkeys originated in Africa, but they had been in contact with numerous species of other animals in transit to Europe. The disease was not known in Uganda, and there was no convincing evidence that the infection was present among wild primates in Central or South Africa.[50-52] No further outbreaks occurred until 1975 when there was one fatal primary case, followed by two secondary cases, in humans in South Africa. The infection had probably been acquired in Rhodesia while the first patient was travelling there, but no vector or animal reservoir could be identified.[53]

Marburg virus disease is almost invariably fatal in monkeys. It is therefore unlikely that the agent is a true simian virus, and it will probably not be found in animals that have been quarantined for longer than 10 days.

In the search for the etiological agent of the first outbreak in 1967, specimens from patients were inoculated by several laboratories into a variety of cell cultures without the production of an obvious gross cytopathic effect.[54,55] In our hands, stained coverslip cultures of primary vervet monkey kidney cells that had been inoculated with serum from a patient in Germany revealed the presence of numerous eosinophilic cytoplasmic inclusions, and we were able to warn other workers that cultures which at first sight might appear to be negative may be found to be positive when stained.[56] By this technique we also detected propagation of the virus in six standard cell lines.[57]

Figure 21 illustrates the appearance of a stained culture of primary vervet kidney cells 10 days after inoculation with Marburg virus, as seen under the low power of the light microscope. The cells had metabolized well, and the sheet was intact. On higher magnification (Figure 22) it can be seen that most of the cells in this culture contain cytoplasmic inclusions while the nuclei remain normal. Our previous experience with simian viruses[58] enabled us to exclude this as a contaminant, and subsequent isolations of the virus from specimens taken from other German patients, together with confirmatory serological tests, established the identity of the agent as the cause of the disease.

C. Ebola Virus

In 1976 there occurred an extensive outbreak of severe hemorrhagic illness in southern Sudan and northern Zaire in Central Africa.[59] A virus morphologically similar to but antigenically distinct from Marburg virus was isolated from patients.[60] This agent has been called Ebola virus, after the Ebola river in Zaire where the first isolate was obtained.[61] The virus has been propagated in the Vero line of vervet monkey kidney cells, with the production of focal rounding of cells.[61-63] Eosinophilic cytoplasmic inclusions have been noted in histological sections of infected liver cells, and it is probable that similar inclusions will be found in stained cell cultures.

III. REOVIRIDAE

The family Reoviridae comprises three genera, namely, *reovirus, rotavirus,* and *orbivirus.*

A. Reovirus

In 1959 Sabin[64] proposed the name reovirus for a group of agents that had been isolated in cell cultures since 1955 from human and nonhuman primate sources. The human prototype strain was designated ECHO virus type 10, but it was clear that it differed both in size and in cultural characteristics from the other ECHO viruses. This strain was renamed Reovirus type 1, and its removal from the echovirus series has left echovirus type 10 category vacant.

Reoviruses can be cultivated in a wide range of cell types.[65] Replication takes place exclusively in the cytoplasm and can proceed in enucleated cells although with less efficiency than in cells with nuclei.[30] In most instances the nuclei of infected cells remain unchanged in appearance until the whole cell degenerates, but occasionally small eosinophilic areas can be noted in the nucleus. At low multiplicity of infection, reovirus has a long maturation time of about 39 hr,[66] and in order to detect small amounts of virus it is necessary to maintain cultures for up to 3 weeks, even when roller tubes which enhance the growth of reovirus[67] are used.

The first description of the eosinophilic cytoplasmic inclusions formed in reovirus-infected cells was provided by us in our report on the cytopathic effects of seven vervet monkey viruses.[68] The inclusions arise first as small well-defined bodies that appear to increase in size and later coalesce to form the characteristic inclusion which occupies the greater part of the cytoplasm and may overlie the nucleus (Figure 23). The inclusion has a smooth texture and does not contain distinct loculations. It stains red with the hematoxylin-eosin method and blue with Giemsa stain (Figure 24), but with May-Grunwald-Giemsa technique it takes on an intermediate color (Figure 25). The genome of the virus consists of double-stranded RNA together with some single-stranded RNA and protein.[69] Virus particles in crystalline array have been observed inside the inclusions.[70,71] With the immunofluorescence technique, the cytoplasmic inclusion fluoresces brightly around the unaffected nucleus.

The reovirus inclusion is morphologically and tinctorially so distinctive that in its fully developed form it can be taken as pathognomonic of reovirus infection. It is a valuable marker when reovirus is present in small amounts only, or when more than one virus is present (Figures 89 and 90). Reovirus occurs in sewage and abattoir effluents,[72,73] and studies on these fluids should include the use of stained cultures.

B. Rotavirus

The name rotavirus is of recent origin, having been proposed in 1974 by Flewett and his colleagues.[74] By means of electron microscopy Bishop et al.[75] had detected particles resembling orbivirus in the cytoplasm of epithelial cells in duodenal biopsy specimens taken from children with acute nonbacterial gastroenteritis. Further electron microscopic studies of stools from gastroenteritis patients confirmed the etiological association between the orbivirus-like particles and the disease. Flewett et al.[76] noted that the agent morphologically resembled the orbivirus of bluetongue disease of sheep rather than reovirus, and observed the similarity between the human agent and the virus of gastroenteritis in calves described by Fernelius et al.[77] They concluded that these viruses might be members of a new group of double-stranded RNA agents.

Studies of agents now known as rotaviruses date back to 1947 when Cheever and Mueller[78] together with Pappenheimer and Enders[79] described diarrheal epizootics that had repeatedly occurred in their mouse breeding colonies. These might have been due to two different viruses: one producing nuclear inclusions, and the other causing cytoplasmic inclusions in epithelial cells of the gut. Further studies by Pappenheimer and Cheever[80] on infected suckling mice confirmed the close association between the formation of cytoplasmic inclusions and the diarrheal disease. It was noted that the inclusion-bearing cells occurred only at the summits of the villi and not in the crypts, and that they desquamated after the first few days of the disease.

Investigation of this epidemic diarrhea of infant mice was continued in 1957 by Kraft[81] who proved that air transmission of the infection occurred and who applied the acronym EDIM to the disease.[82] Kraft also noted the presence of antibody to the mouse agent in human sera. The virus was then studied by electron microscopy,[83,84] and it was observed that morphologically the EDIM agent resembled reovirus. Rubenstein et al.[85] obtained some evidence of growth of the virus in organ cultures of mouse intestine. Woode and his colleagues[86] showed by immunofluorescence that the virus infected primary cultures of mouse kidney and human embryo kidney cells.

In 1963 we described[58] unusual cytoplasmic inclusions produced by a virus isolated 5 years previously from the rectum of a healthy vervet monkey (Figure 26), and we designated the agent SA 11 in the series of vervet monkey viruses. In 1967 we isolated a similar agent on three occasions from the effluent of an abattoir, and we called it the O agent because it was recovered from the washings of offal from cattle or sheep.[73] The presence of the virus was only detected because we stained all inoculated cultures and were thus able to discern inclusions in the absence of a gross cytopathic effect. We then compared SA 11 and the O agent further,[87] and established their close relationship. Electron microscopic studies by Lecatsas and Els[88,89] indicated the morphological resemblance between these two viruses and the reovirus-like agent of human gastroenteritis. Further comparison between the human and animal rotaviruses by Kapikian et al.[90] has led to the use in many laboratories of SA 11 as a substitute antigen in tests for the detection of human rotavirus infection.

Mebus and his colleagues in 1971 reported[91] the propagation of neonatal calf diarrhea (NCD) virus in primary fetal bovine cell culture. Two strains of the virus were cultivated: the Cody strain which grew to a limited extent in primary fetal bovine lung cells, and the Lincoln strain which preferred fetal bovine kidney cells and was carried successfully through 37 passages. A grossly detectable focal degeneration, progressing to complete destruction of the cell sheet, occurred; and the presence of virus in the cytoplasm was demonstrated by immunofluorescence. Further studies on these two strains of Nebraska calf diarrhea virus were reported by Fernelius et al.[77] who found that the virus had a reovirus-like morphology but was not serologically related to two reoviruses tested. Welch and Twiehaus[92] stained primary fetal bovine kidney cells infected with both strains of NCDV agent, and observed that cytoplasmic vacuolation developed within 24 hr, while a few cells contained eosinophilic cytoplasmic inclusions a few days after infection.

We have cultured the Nebraska agent in the MA104 line of fetal rhesus kidney cells, with good production of typical rotavirus inclusions. Vacuolation of the cytoplasm has not been a notable feature of the infection in this type of cell. The inclusions are multiple and discrete (Figures 27 and 28). In this cell line both the calf and the simian agents produce virtually identical cytopathic changes. The nucleus usually remains unchanged until the whole cell degenerates, when the nucleus shrinks and becomes pyknotic. Occasionally small eosinophilic areas appear in the nucleoplasm. The central portion of the cytoplasm is frequently more eosinophilic than the outer part. Small loculations may be seen in the larger inclusions, but they do not contain basophilic material. The infected cell eventually shrinks and stains a deeper red, and the presence of such eosinophilic small cells scattered throughout the culture is characteristic of infection with these rotaviruses. A comparison between SA 11 inclusions stained with Giemsa, May-Grunwald-Giemsa, and hematoxylin-eosin stains is shown in Figures 29 to 31.

C. Orbivirus

With the exception of Colorado Tick Fever virus, which causes a mild febrile illness in man, the arthropodborne orbiviruses are not of significance in human disease as far as is known. In veterinary medicine, on the other hand, the orbiviruses causing African horse-sickness and bluetongue of sheep are of considerable importance.

It is generally accepted that, like other reoviridae, the orbiviruses replicate in the cytoplasm. In several studies by Ozawa and his colleagues[93-97] monkey kidney cells infected with African horse-sickness virus showed nuclear changes, including some suggestive of adeno- and papovaviruses, while virus particles were observed in the nucleus. Lecatsas and Erasmus[98] were unable to confirm these findings and clearly showed by electron microscopy that the cytoplasm alone is the site of virus replication.

Bluetongue virus has been cultivated in a variety of cell cultures,[99-104] and in most instances presence of the virus has been demonstrated by a gross cytopathic effect, by electron microscopy or by immunofluorescence. The agent produces basophilic cytoplasmic inclusions which differ in appearance from those caused by reo- and rotaviruses. Through the courtesy of Dr. Stewart McConnell of Texas A & M University the appearance of bluetongue inclusions is illustrated in Figure 32.

IV. POXVIRUSES

The poxviruses all have the following characteristics: they are DNA viruses with a comparable complex morphology; they possess a common internal nucleoprotein antigen but vary in their external antigens; and all replicate in the cytoplasm alone, usually with the production of cytoplasmic inclusions.

Poxviruses are classified in groups chiefly according to the main animal host and the closeness of serological relationships. The most important group in human disease is that of the variola-vaccinia viruses, which includes the agents of smallpox (variola), vaccinia, cowpox, rabbitpox, monkeypox, and mousepox (infectious ectromelia). The apparently successful eradication of smallpox disease in recent years is a remarkable achievement, but for the present some questions remain unanswerable. Variola virus still exists in a few recognized laboratories, and probably also in some unrecognized laboratories. As a potential means of microbiological warfare by the vaccinated against the unvaccinated, variola virus has acquired a special significance since the routine vaccination of infants has been discontinued in a number of countries. Furthermore, one closely related poxvirus of animals can infect man, and in view of the recognized capacity of poxviruses to form variants, there can be no absolute assurance that a virus with the ability to cause serious human disease will not arise in the future.

Agents of the variola-vaccinia group can be cultivated in a variety of cells, including some of human, monkey, chick, and mouse origin. Gross cellular changes can be seen as focal rounding of cells with or without syncytium formation.

It has long been recognized that two kinds of cytoplasmic inclusions are formed which differ morphologically and tinctorially. The use of the terms A type and B type for these can lead to confusion with the entirely different Cowdry type A and type B nuclear inclusions. The A type cytoplasmic inclusions of poxviruses are generally taken to be somewhat indistinct bodies that are Feulgen-negative for DNA and usually do not contain virus particles. Ichihashi and Matsumoto[105] have, however, divided the A type inclusions of mousepox and cowpox into two kinds: one that contains mature virions, and one in which there are no virions. Kato et al.[106] also observed virus particles in the A type inclusions of cowpox virus. The B type inclusions are considered to be more granular and well defined. They are Feulgen-positive for DNA and usually contain virus particles.

Poxvirus inclusions are often described as being eosinophilic (acidophilic), although a variety of tinctorial properties has been ascribed to them. Kato et al.[107] cultivated several viruses of the variola-vaccinia group in human FL and HeLa cells, and found that with methanol fixation followed by Giemsa staining, most of the inclusions appeared reddish-purple and corresponded to B type inclusions; whereas cowpox virus, which usually produces chiefly the A type, formed inclusions that stained pale blue. When infected cells were fixed with Bouin's fluid and stained with hematoxylin and eosin, the tinctorial properties of both A and B type inclusions changed completely: the B type were basophilic with only an eosinophilic tinge, while the A type were brightly eosinophilic. This distinction between the basophilic B type and the acidophilic A type inclusions is illustrated in Figure 33 which shows both types produced in monkey kidney cells by a freshly isolated strain of variola minor virus. Kato and his colleagues quoted Guarnieri as describing the inclusions of variola and vaccinia as "basophil, hematoxylinophil". They considered that the term Guarnieri body should, strictly speaking, be applied only to inclusions produced by variola and vaccinia viruses, and that similar inclusions formed by other poxviruses should be called B type.

Kato and Cutting[108] listed the various staining reactions ascribed in the literature to the inclusions produced by different poxviruses. They observed that with methanol fixation no halo was seen around inclusions, whereas Bouin's and Zenker's fluids caused shrinkage and halo formation and defined the A type inclusions more clearly.

When Bouin's fixative is used in combination with hematoxylin and eosin, the basophilic B type inclusions of various poxviruses are seen to be granular and may show loculations containing basophilic internal bodies (Figure 34) which are virtually pathognomonic of poxvirus infection. Syncytium formation (Figure 36) by the variola-vaccinia group depends on both the strain of virus and the type of cell used for its cultivation.[107, 109-113] Variola virus (Orthopox variolae[114]) can be propagated in a variety of cell types.[115] Figures 33 to 36 illustrate the range of cytopathic changes seen with hematoxylin and eosin staining: in the same preparation some inclusions stain faintly while others are more deeply basophilic. Vaccinia virus (Orthopox officinale; Orthopox commune) can also be grown in a wide range of cells with the production of cytoplasmic inclusions and, with some strains, syncytia. Figure 37 shows the basophilic inclusions produced in Vero cells by vaccinia virus.

Vaccinia virus today comprises a number of strains whose origins are obscure. Jenner himself appears to have used a horse poxvirus as well as cowpox virus for human vaccination,[116] and the extant vaccinia strains may have derived from equine, bovine, or human poxviruses. Although poxviruses replicate solely in the cytoplasm, it has been shown by Hruby et al.[117] that in enucleated cells vaccinia virions do not complete their maturation, indicating that some nuclear function is necessary for the replicative cycle.

Mousepox was first described and named infectious ectromelia in 1930 by Marchal[118] who noted the cytoplasmic inclusions in histological sections of infected mouse tissues. The agent was propagated in cultures of mouse and chick tissue fragments by Downie and McGaughey[119] in 1935. The inclusions were further investigated by Ichihashi and Matsumoto[120] and by Kamahora et al.[121] The latter workers concluded that one infectious particle is sufficient to establish one B type inclusion in a cell. Giant cell formation by mousepox virus was noted by Nii[122] and by Barski.[109]

Monkeypox virus also produces loculated basophilic cytoplasmic inclusions (Figure 38) and can be cultivated in a wide range of cell types.[123-125] There have been 36 human cases of monkeypox reported from Africa since the first case in 1970.[126-129] The area involved is very large, extending from Zaire to Sierra Leone, and probably there have been more cases than those reported. The disease resembles smallpox clinically and has had a case fatality rate of 16.6% to date. Transmission of monkeypox to contacts of human cases has been minimal. Man appears to be relatively insusceptible, as there were no human cases associated with a number of outbreaks of monkeypox in laboratory primates in Europe and the U.S.[130] There is no evidence that monkeys can serve as a reservoir of variola virus, and it seems unlikely that monkeypox virus in its present form will cause extensive epidemics in man, but there is always the possibility that a mutant strain with epidemic propensity may occur in the future. It should be borne in mind that at least two other variola-related poxviruses have been isolated in Africa.[131, 132]

Another simian poxvirus, Tanapox, can infect man naturally. It does not belong to the variola-vaccinia group, but is serologically related to the Yaba virus of African monkeys. Yaba virus was first reported in 1958 from Nigeria, where subcutaneous tumors occurred in imported rhesus monkeys housed in cages adjacent to those of African monkeys.[133] The infection appears to be enzootic in African monkeys in a subclinical form. Yaba virus has been cultivated in monkey kidney cells with the development of focal lesions and the production of cytoplasmic inclusions.[134-137]

Tanapox disease in humans was first noted in 1957 among African schoolchildren near the Tana River in Kenya. A second outbreak further up the river occurred in 1962, probably involving several hundreds of persons of all ages. The illness was of short duration, with headache, prostration, and usually only a single pocklike lesion on the upper part of the body. Both outbreaks happened when the river was in flood and the population took refuge on islands where there was unusually close contact with wild and domestic animals. A skin biopsy from a patient in the second outbreak yielded a poxvirus unrelated to the variola group.[138] The infection subsequently occurred in monkeys at three primate centers in the U.S. between 1965 and 1968. The poxvirus recovered from these outbreaks was called Yaba-like or OrTeCa (an acronym for Oregon, Texas, and California where the primate centers were situated). Several animal handlers contracted the infection through small scratch wounds on their arms.[139] The virus was identical to that which caused the 1962 outbreak in Kenya, and Downie and Espana[140] have proposed that only the name Tanapox be used for it. Antibody against Tanapox virus has been found in monkeys, baboons, and chimpanzees from Central Africa in areas as widely separated as Ethiopia and Senegal.[141, 142]

Casey et al.[143] studied histological biopsy specimens taken from lesions in monkeys at the Texas primate center, by both electron and light microscopy. After fixation with formalin and staining with hematoxylin and eosin, infected cells showed granular, irregularly shaped cytoplasmic inclusions, and vacuoles containing amorphous pink-staining material were seen in the nuclei. It was noted that swinepox virus also produces intranuclear vacuoles but that ultrastructural studies had shown that these are composed of fine fibrils.

Figures 39 and 40 illustrate the changes induced by Tanapox virus in cell culture.

Nuclear vacuoles containing faintly eosinophilic material are a prominent feature. The predominantly basophilic cytoplasmic inclusions are coarsely granular and contain internal structures.

The virus isolated from monkeys in Oregon was cultivated in a variety of cell types by Nicholas and McNulty,[144, 145] who also observed cytoplasmic inclusions and nuclear vacuoles which, they noted, resembled those produced by swinepox, sheeppox, and myxoma viruses. For further studies on the cytopathic changes produced by swinepox virus, reference may be made to papers by de Boer and Teppema.[146, 147]

V. ARENAVIRUSES

The name arenavirus,[148] or arenovirus,[149] derives from the sandy appearance of the virion in which are incorporated electron-dense granules resembling ribosomes of the host cell. Lymphocytic Choriomeningitis (LCM) virus is the prototype of this group of RNA agents which includes the Tacaribe complex of viruses (some of which are associated with South American hemorrhagic fevers) as well as Lassa virus.

The arenaviruses were previously classified as arboviruses, but spread to man at least does not require a vector. Natural reservoirs include wild rodents and the house mouse, in which persistent infections are established. Transmission to man is chiefly by way of rodent excreta.

The propagation of LCM virus in cell cultures has been reviewed by Lehmann-Grube,[150] who has commented that the virus characteristically leaves intact the cells in which it multiplies, although a few exceptions have been reported. Immunofluorescence studies indicate that replication as a rule takes place only in the cytoplasm. Benda and Cinatl[151] cultivated LCM virus in various human and animal cell cultures without the production of gross cytopathic changes, and they were unable to confirm earlier reports of overt cytopathic effects in KB cells or primary mouse and chick cells.

By means of a technique developed by Ackermann and cited by Lehmann-Grube,[150] LCM virus grown in cell culture under an agar overlay produces plaques of cells which fail to stain with neutral red but show no other visible alteration. On the other hand, the other arenaviruses produce normal type plaques under agar overlay.[152]

Mettler et al.[153] inoculated Junin virus, a member of the Tacaribe complex, into HeLa cells and obtained gross focal changes, with rounding and fallout of cells, proceeding to total involvement of the cell sheet. Giemsa staining showed the presence of polymorphic basophilic cytoplasmic inclusions. Buckley[154] similarly cultivated Tacaribe virus in HeLa cells with the same results. Immunofluorescence showed cytoplasmic localization of antigen in sites corresponding to the inclusions seen by light microscopy.

Seven viruses of the Tacaribe complex were inoculated into Vero cells by Murphy et al.[155] for electron microscopy. Each virus produced in the cytoplasm electron-dense matrices in which were dispersed masses of ribosome-like granules identical to those present in virions. These matrices corresponded in size and shape to the inclusions seen under the light microscope. Maturation of virus particles occurred at the cell periphery where virions budded from the plasma membrane.

Lassa fever was first encountered in 1969 in Nigeria, involving three nurses of whom two died. In the following year there was a second outbreak affecting 28 persons, with 13 deaths. Further outbreaks took place in Liberia and Sierra Leone. The overall case fatality rate has been approximately 43%. The causative virus was isolated in Vero cells by Buckley et al.[156] Gross focal cytopathic changes were noted a few days after inoculation, and these progressed to complete destruction of the cell sheet within about 1 week. Cellular destruction was accompanied by increased acid production in the medium, changing the phenol red indicator from red to yellow. Staining with Giemsa showed the presence of basophilic pleomorphic cytoplasmic inclusions resembling those produced by Junin virus.[153] Electron microscopy of infected Vero cells[157] indicated that Lassa virus morphologically belongs to the same taxon as LCM and the Tacaribe viruses.

Through the courtesy of Dr. Sonja M. Buckley we illustrate in Figure 41 the blue-stained irregularly shaped inclusions produced by Lassa virus in Vero cells, fixed with Bouin's fluid and stained with Giemsa.

VI. RETROVIRUSES

The family of Retroviridae[158] (or Retraviridae[159]) comprises the spumaviruses, lentiviruses, and oncoviruses. All of these three subfamilies can cause the formation of syncytia without inclusions.

Syncytium formation is a normal physiological event in the trophoblast layer of cells around the developing embryo. It can also be induced in other cells by a wide range of viruses, usually in combination with inclusions in the nucleus or cytoplasm. The retroviruses form syncytia without inclusions, and it is important to stain cultures infected with potential candidate viruses to note whether inclusions are present. Failure to do so can lead to incorrect preliminary assignment of agents to this group. The distinction between syncytium formation and cytophagocytosis has already been discussed in Chapter 2, IV.

A. Spumaviruses (Foamy or Syncytial Viruses)

Spumaviruses derive their name from the cytoplasmic vacuolation often found in infected cells. This may vary from relatively small vacuoles (Figure 42) to large foamy masses (Figure 43). This feature is not constantly present, and syncytium formation without nuclear or cytoplasmic inclusions is the only consistently characteristic change (Figure 44). The nuclei initially remain unchanged in appearance but later undergo pyknosis or clumping of chromatin.

The cytopathic effects produced by foamy viruses were first noted by Enders and Peebles[160] in 1954 when latent simian agents appeared in their cultures, and further instances of indigenous foamy viruses present in primary monkey kidney cultures were reported in 1955 by Rustigian et al.[161] Since then at least eight serotypes of simian foamy virus have been identified, as well as similar agents in other animal species including chicken, cat, hamster, and bovine viruses.[162-164]

In 1968 Hsiung[165] reported the isolation of two foamy viruses from human kidney cell cultures prepared over a 3-year period. Another syncytial agent was recovered from lymphoblastoid cells originating from a nasopharyngeal carcinoma in an African patient.[166,167] Comparison of this virus with syncytial agents from other species showed that it was closely related to the chimpanzee simian foamy virus type 6.[168,169] A different foamy agent isolated from the peripheral blood cells of a patient with chronic myeloid leukemia was found to be serologically related to simian foamy virus type 1.[170] Cameron et al.[171] isolated and reisolated a foamy virus from a specimen of brain taken from a patient who had been on renal dialysis for 20 months and who had died of dialysis dementia and progressive encephalopathy.

In view of the number of foamy viruses recovered from animals, further human strains may well be expected to occur. Most of the simian foamy viruses have been isolated from primary kidney cell cultures, and more experience with the cultivation of human kidney tissue beyond the fetal or neonatal stages may be necessary to reveal whether human foamy viruses adopt the same habit of establishing latent infections in kidney cells. Inoculation of urine samples will probably yield far less information than will the cultivation of kidney cells. The simian foamy virus types 6 and 7 were isolated from nervous tissues taken from chimpanzees as well as from their viscera and blood leukocytes;[164] and the human agents recovered from blood cells and brain may indicate the widespread distribution foamy viruses can achieve in the body.

Cells infected with simian foamy agents tolerate simultaneous infection with other viruses, as will be further discussed in a later section. When primary simian cells are used, at least 20% of the growth batch should be set aside as uninoculated control tissue if it is necessary that indigenous viruses should be detected. If syncytial changes occur in inoculated primary simian cell cultures, special attention should be paid to the possibility that foamy virus might be the cause of such cytopathic change.

B. Lentiviruses

The lentiviruses comprise the causative agents of two diseases of sheep in Iceland and possibly elsewhere as well.[172,173] Visna, meaning shrinkage or wasting, involves the central nervous system with demyelination and paralysis; Maedi, meaning dyspnoea, is a chronic progressive pneumonia.[174] These two RNA viruses are closely related but show only a partial cross-reaction in the neutralization test.[175] The long incubation period of months or years gives the name lentivirus to these slow viruses.

Inoculation of Visna and Maedi virus-infected tissue suspensions into cultures of sheep choroid plexus cells results in the formation of syncytia within 2 or 3 weeks. No nuclear or cytoplasmic inclusions are present in the syncytia.[176,177] Thormar and Sigurdardottir[178] inoculated Visna virus into human embryonic and 1- to 4-day-old calf, pig, cat, dog, and guinea pig choroid plexus cells, as well as into human and calf kidney cells. Clear evidence of viral replication and syncytium formation was obtained only with calf cells.

Haase et al.[179] studied the question of why infectious Visna virus could not readily be detected in the tissues of infected animals, whereas the virus appeared within a few days in explants cultured from the same animals. There was evidence of nuclear localization of Visna provirus, and it was concluded that in vivo there is a block in virus replication at the provirus stage except in a small proportion of cells, which could account for the slow spread of virus. In vitro, on the other hand, spontaneous activation of the virus takes place in a larger proportion of cells.

Visna virus was inoculated by Macintyre et al.[180] in parallel into sheep choroid plexus cells and into a line of human astrocytes derived from a malignant tumor. Syncytium formation followed by complete destruction of the culture occurred in the sheep cells. The human astrocytes were not killed by the virus although large syncytia were formed. If the sheet was allowed to grow until it rolled off the surface of the plastic flask, the sparse remaining cells grew out and repeated the cytopathic pattern. This cycle was maintained for 4 to 5 months.

August and Harter[181] inoculated Visna virus into four monkey kidney cell lines: BSC-1, Vero, LLC-MK2, and CV-1. No significant propagation of virus occurred, but syncytia formed within 24 hr after infection, especially in the BSC-1 and Vero cells, and cytoplasmic vacuolation was common in the syncytia. It was considered that the two most susceptible cell lines could serve as indicator systems for Visna virus plaque assay.

C. Oncoviruses

The oncoviruses[182] include agents previously called oncornaviruses or leukoviruses, which cause leukemias or sarcomas in animals. Oncoviruses are grouped as A-, B-, C-, or D-type particles chiefly on morphological grounds, and they are also distinguished according to whether they can replicate in cells of the natural host (ecotropic), in cells only of a different species (xenotropic), or in cells of both the natural host and of a different species (amphotropic).

These RNA agents, by means of their reverse transcriptase enzyme, produce DNA provirus which can be integrated into the genome of the host cell. Such an incorporated provirus can be genetically transmitted vertically and may be present in cultured cells. Replication takes place when certain physical or chemical inducers are applied to the culture, or when the cells are co-cultivated with permissive cells in which viral expression can occur.

Klement et al.[183] introduced the mixed cell cytopathogenicity test: mouse embryo cells infected with mouse leukemia virus showed no overt changes, but when they were co-cultivated with rat tumor cells (XC cells) that had been transformed by the avian Rous sarcoma virus, the XC rat cells fused into syncytia. The test has been extended to include the detection of oncoviruses of other species[184] and the use of other cells as the indicator system.[185-188] Foam has been present in syncytia in some of these tests.

Ample evidence of the latent or productive presence of oncoviruses in established cell lines (including human cell lines[189]) has been presented by Lieber et al.[190] and by a number of other workers; and Weiss[191] has warned of the pitfalls that may be encountered when oncoviruses are present in cell cultures used for experiments with exogenous viruses.

VII. CORONAVIRUSES

Some strains of both human and animal coronaviruses produce syncytia in cell cultures; others cause degeneration of individual cells, resulting in rounding, vacuolation, or a stringy appearance of the cells. Replication of virus takes place exclusively in the cytoplasm, but staining of infected cells has shown that visible inclusions are not present.[192] References to the cultivation of coronaviruses in various cell cultures can be found in the comparative review of coronaviruses by McIntosh.[193]

VIII. RESPIRATORY SYNCYTIAL VIRUS (PNEUMOVIRUS)

The respiratory syncytial virus of man and chimpanzees together with the pneumonia virus of mice are at present classified as the genus *Pneumovirus* in the family of Paramyxoviridae.

In 1956 Morris et al.[194] isolated a virus from chimpanzees with an illness characterized by coughing, sneezing, and nasal discharge. The agent was shown by serological tests to have been the presumptive cause of the illness, and it was called the chimpanzee coryza agent (CCA). It produced degeneration in the Chang strain of human liver cells as well as in monkey kidney cells, about 1 week after inoculation of the cultures. Cytopathic changes were difficult to interpret because nuclear and cytoplasmic inclusions occurred in both the uninoculated and the inoculated liver cells, and the monkey kidney cells contained adventitious simian agents. Several other types of cells were tried but proved to be less susceptible to the coryza agent.

In the following year Chanock et al.[195] reported the recovery of two syncytium-producing viruses from humans: one from an infant with bronchopneumonia, and the other from an infant with laryngotracheobronchitis. These viruses were compared with other syncytium-forming agents, including the chimpanzee coryza agent, and they were both found to be indistinguishable from the CCA. The strains of the human virus grew well in the KB line of human epidermoid carcinoma cells and in Chang liver cells, but growth was slower in human amnion cell cultures. In KB cells stained with May-Grunwald-Giemsa, nuclei in syncytia contained a dense central mass resembling an inclusion from which thick chromatin strands radiated. The virus possessed no hemagglutinin for chicken or human O erythrocytes, was not pathogenic for newborn mice, and did not grow in embryonated eggs. Chanock and Finberg[196] proposed that the two agents isolated from human infants be grouped together with the chimpanzee coryza agent and be called Respiratory Syncytial Virus.

Jordan[197] studied differences in syncytium production by one of the human strains in various cells and nutrient media. Syncytia were more marked in HEp-2 cells than in KB or HeLa cells, and Eagle's MEM medium appeared to have some unspecifiable factors that favored syncytium formation. Alkalinity of the medium to some extent increased the development of syncytia. Virus production was not in direct proportion to the amount of syncytium formation, and about half of the virus produced was cell-bound when cytopathic changes were maximal.

Armstrong et al.[198] compared electron and light microscopic preparations of infected HeLa cells. Electron microscopy showed that in the syncytia formed by respiratory syncytial virus, cytoplasmic structures corresponding in distribution to the eosinophilic inclusions seen by the light microscope were of two kinds. One kind consisted of a fine fibrillar matrix, and the other was composed of denser amorphous or granular material. Virus particles appeared to be produced virtually entirely by single spindle-shaped cells and not by the syncytia. Virions budded at the cell surface and were never observed within the main body of the cell. In the single spindle-shaped cells only the fibrillar matrix type of inclusion was seen. Electron microscopy also revealed that in the nucleus there were finely fibrillar areas similar to those present in the cytoplasm, but no nuclear inclusions were seen under the light microscope. The nucleoli were enlarged and bizarre in shape. Armstrong and his colleagues came to the conclusion that synthesis of new virus preceded the formation of syncytia and discrete inclusions.

Norrby et al.[199] noted that the nuclei of infected Vero monkey kidney line cells showed no significant changes by electron microscopy until the cells started to become pyknotic. Syncytium formation was not extensive in Vero cells. Compact cytoplasmic inclusions were numerous in the late stages of virus development and had a granular or threadlike composition. In contrast, it was noted, the cytoplasmic inclusions formed by other paramyxoviridae are less dense and contain loosely arranged nucleocapsids having a diameter of 17 to 18 nm.

The cytochemistry of the inclusion bodies formed by respiratory syncytial virus in HEp-2 and U-amnion cells was examined by Corbitt and Beswick.[200] Methyl green-pyronin stained the inclusions deep pink, suggesting the presence of RNA; but with acridine orange the inclusions fluoresced green, possibly due to nucleoprotein binding the dye and thus masking the presence of nucleic acid. A similar effect has been noted in reovirus inclusions by Love and Fernandes.[69]

Gharpure et al.[201] grew respiratory syncytial virus in the presence of the chemical mutagen 5-fluorouridine and produced a temperature-sensitive mutant designated ts2 which did not induce the formation of syncytia. Belshe et al.[202] concluded that this mutant probably has a single genetic lesion that is responsible for both the temperature sensitivity and the lack of a protein factor for cell fusion. In Figure 45 we illustrate syncytium and cytoplasmic inclusion formation produced by respiratory syncytial virus in Vero cells.

The basic principles involved in the fusion of cell membranes have been extensively discussed by Poste and Allison.[203] Fusion of intracellular membranes occurs in the course of normal processes such as digestion and secretion. Fusion of the outer cytoplasmic membranes of two or more cells is a normal event in the formation of the syncytial trophoblast of the embryo, but it occurs pathologically in response to chronic inflammation, in some tumors, and as the result of the action of certain viruses. The mechanisms of virus-induced cell fusion have been fully considered by Poste[204, 205] who has listed the viruses causing cell fusion, together with publication references for each. Bratt and Gallaher[206] distinguished between the two types of virus-induced cell fusion: fusion from without, and fusion from within. The former is essentially a laboratory artifact produced in cell cultures either by viral components acting independently of the infectious virion, as in the case of inactivated Sendai virus; or by very large doses of infectious virus applied to the cells, in which case cell fusion occurs long before the end of the latent period and does not require intracellular protein synthesis. Several RNA agents, including the paramyxoviridae and Visna virus, as well as two DNA agents, *herpesvirus hominis* and vaccinia virus, have been shown to cause cell fusion from without when administered to cell cultures in large doses. Fusion from within, which follows penetration of the cell by virus applied in small doses, requires intracellular protein synthesis but not necessarily synthesis of complete virions. It can be produced by defective virus undergoing an abortive cycle, as well as by infectious virus performing a complete replication cycle.

IX. OTHER PARAMYXOVIRUSES

Other paramyxoviruses, including Newcastle disease virus, mumps, parainfluenza types 1, 2, and 4, and the simian agent SV 5 produce irregularly shaped eosinophilic cytoplasmic inclusions, but nuclear inclusions are not, as a rule, present. Syncytia are frequently formed by these viruses. It is generally accepted that replication occurs only in the cytoplasm. Parainfluenza type 3, which produces nuclear inclusions as well as cytoplasmic inclusions, will be considered in Chapter 7, Section II.

A. Newcastle Disease Virus

Newcastle disease virus is an avian agent of considerable importance in the poultry industry. It is the type species virus in the *paramyxovirus* genus.[207] Brandt[208, 209] described the cytoplasmic inclusions produced in chick embryo and various mammalian cells, noting that they could not be detected unless acidic fixatives like Bouin's or Zenker's fluids were used prior to staining with hematoxylin and eosin. The nucleus remains unaltered morphologically, except under special conditions such as the treatment of Newcastle disease virus-infected hamster kidney cells with plant lectins, when marked nuclear degeneration results.[210]

B. Mumps Virus

The cytoplasmic inclusions of mumps virus were first described by Brandt.[208] Gresser and Enders[211] confirmed that nuclear inclusions were not seen in primary cultures of chick embryo cells and human amnion cells infected with mumps virus. They noted that a chick-adapted strain did not grow or produce cytopathic changes in human amnion cells, and an amnion-adapted strain did not grow in chick cells. In the homologous cells each strain produced syncytia and cytoplasmic inclusions. In the human amnion cultures both multinucleated and spindle-shaped cells were induced, but attempts to isolate virus variants that would produce only one or the other type of cellular change were unsatisfactory. It was noted that hemadsorption with guinea pig erythrocytes was positive before gross cytopathic changes were seen in the cultures. In the appropriate cells, plaques under agar overlay could be produced.

Nasibov and Smorodintsev[212] reported an inverse relationship between virus yield and inclusion production in various cell types following infection with a standard dose of mumps virus. Using a combination of methylgreen-pyronin stain and fluorescein-labeled antibody, they detected viral changes in the nucleus as well as in the cytoplasm, but specific viral antigen was not present in the structures described as inclusions in either site. Other immunofluorescence studies[213, 214] have also shown that mumps antigen is not found in the nucleus, but the observation by Nasibov and Smorodintsev that morphological changes may occur in the nucleus has not been in keeping with the findings of most other investigators. Nevertheless, some viruses induce nuclear inclusions that do not contain viral antigens, and it is possible that strains of mumps virus may vary in this respect. For instance, we isolated a virus from the mouth of a clinical case of mumps infection in primary human amnion cells. No syncytia were produced, and at the 17th day only about 2% of the cells contained either cytoplasmic or nuclear inclusions or both. On passage from amnion into vervet monkey kidney cells very few small syncytia were seen, but by the 21st day two thirds of the cells had cytoplasmic inclusions, and one third had both nuclear and cytoplasmic inclusions. Figure 85 illustrates the type of inclusions observed.

C. Parainfluenza Virus

Parainfluenza virus types 1, 2, and 4 produce cytoplasmic inclusions that are sometimes accompanied by syncytia (Figure 82). In contrast to respiratory syncytial virus, which tends to produce inclusions in single cells only in close proximity to syncytia, the parainfluenza viruses in vervet monkey kidney cells induce inclusions in single cells throughout the culture. In the absence of syncytia the effect closely resembles that of Marburg virus.

Various cell types differ in their susceptibility to the parainfluenza viruses. Primary monkey kidney cell cultures have been widely used for their isolation, and for this purpose the LLC-MK2 line of rhesus kidney cells can be used as an alternative.[215] Serial passage of rhesus monkey kidney cells may, however, lead to failure of their capacity to support the growth of the Sendai strain of parainfluenza type 1, owing to the loss of a trypsin-like proteolytic factor.[216] Primary and secondary human embryo kidney epithelial cells have been found to be sensitive to parainfluenza types 1 and 2,[217] but a strain of human embryo lung cells which supported the growth of type 2 was unsuitable for type 4 parainfluenza.[5] Canchola et al.[218] isolated a number of strains of type 4 more readily in primary rhesus kidney cells than in vervet kidney cells, but all strains were first detected only by hemadsorption 2 to 4 weeks after inoculation of the cultures with throat swab specimens. Sensitivity of the cells was increased tenfold by rolling the cultures.

D. Simian Agent SV 5

The simian agent SV 5 has been encountered frequently as an endogenous contaminant of kidney cultures from several species of monkeys. The virus is sometimes referred to as parainfluenza type 5, and the article by Hsiung[219] should be consulted for a full discussion of the possible role of SV 5 in human infection. A one-way antigenic cross-relationship between SV 5 and parainfluenza type 2 has been demonstrated by complement fixation, but the viruses are distinct by the hemagglutination-inhibition test.[220] SV 5 also has a serological relationship with mumps virus.[221]

Primary rhesus monkey kidney cells support the growth of SV 5 with little grossly visible change except occasional small syncytia, but Choppin[222] showed that infected cultures under agar overlay, stained with neutral red, reveal plaques consisting of morphologically intact cells that take up the stain to a lesser extent than the surrounding cells. In contrast, the BHK-21 line of hamster kidney cells produces relatively little virus but shows large syncytia which contain eosinophilic inclusions in the cytoplasm, corresponding to areas of nucleocapsid aggregation seen by electron microscopy.[223,224] Hemadsorption with guinea pig red cells is commonly used for the detection of SV 5 as a contaminant of monkey kidney cultures in which gross cytopathic changes are not obvious.

REFERENCES

1. **Cooper, P. D., Agol, V. I., Bachrach, H. L., Brown, F., Ghendon, Y., Gibbs, A. J., Gillespie, J. H., Lonberg-Holm, K., Mandel, B., Melnick, J. L., Mohanty, S. B., Povey, R. C., Rueckert, R. R., Schaffer, F. L., and Tyrrell, D. A.**, Picornaviridae: second report, *Intervirology*, 10, 165, 1978.
2. **Tyrrell, D. A. and Parsons, R.**, Some virus isolations from common colds. III. Cytopathic effects in tissue cultures, *Lancet*, 1, 239, 1960.
3. **Hamre, D.**, Critique of methods for recovery of respiratory viruses, *Am. Rev. Respir. Dis.*, 88, 306, 1963.
4. **Stott, E. J. and Tyrrell, D. A.**, Some improved techniques for the study of rhinoviruses using HeLa cells, *Arch. Gesamte Virusforsch.*, 23, 236, 1968.
5. **Votava, M. and Tyrrell, D. A.**, The possible value of certain cells for the propagation of respiratory viruses, *Arch. Gesamte Virusforsch.*, 29, 253, 1970.
6. **Strizova, V., Brown, P. K., Head, B., and Reed, S. E.**, The advantages of HeLa cells for isolation of rhinoviruses, *J. Med. Microbiol.*, 7, 433, 1974.
7. **Hitchcock, G. and Tyrrell, D. A.**, Some virus isolations from common colds. II. Virus interference in tissue cultures, *Lancet*, 1, 237, 1960.
8. **Olson, L. C., Willhight, M., and Buescher, E. L.**, Recovery and characterization of non-cytopathogenic rhinoviruses, *J. Gen. Virol.*, 17, 237, 1972.
9. **Tyrrell, D. A.**, Rhinoviruses, *Virol. Monogr.*, 2, 67, 1968.
10. **Wenner, H. A. and Lenahan, M. F.**, Propagation of group A coxsackie viruses in tissue cultures. II. Some interactions between virus and mammalian cells, *Yale J. Biol. Med.*, 34, 421, 1961/2.
11. **Enders, J. F., Holloway, A., and Grogan, E. A.**, Replication of poliovirus 1 in chick embryo and hamster cells exposed to Sendai virus, *Proc. Nat. Acad. Sci. U.S.A.*, 57, 637, 1967.
12. **Neff, J. M. and Enders, J. F.**, Poliovirus replication and cytopathogenicity in monolayer hamster cell cultures fused with beta propiolactone-inactivated Sendai virus, *Proc. Soc. Exp. Biol. Med.*, 127, 260, 1968.
13. **Neff, J. M. and Enders, J. F.**, Further observations on replication of type 1 poliovirus in naturally resistant fused cell cultures, *Perspect. Virol.*, 6, 39, 1968.
14. **Ackermann, W. W., Rabson, A., and Kurtz, H.**, Growth characteristics of poliomyelitis virus in HeLa cell cultures: lack of parallelism in cellular injury and virus increase, *J. Exp. Med.*, 100, 437, 1954.
15. **Reissig, M., Howes, D. W., and Melnick, J. L.**, Sequence of morphological changes in epithelial cell cultures infected with poliovirus, *J. Exp. Med.*, 104, 289, 1956.
16. **Buckley, S. M.**, Cytopathology of poliomyelitis virus in tissue culture. Fluorescent antibody and tinctorial studies, *Am. J. Pathol.*, 33, 691, 1957.
17. **Shaver, D. N., Barron, A. L., and Karzon, D. T.**, Cytopathology of human enteric viruses in tissue culture, *Am. J. Pathol.*, 34, 943, 1958.
18. **Robbins, F. C., Enders, J. F., and Weller, T. H.**, Cytopathogenic effect of poliomyelitis viruses in vitro on human embryonic tissues, *Proc. Soc. Exp. Biol. Med.*, 75, 370, 1950.
19. **Barski, G., Endo, M., and Monaci, V.**, Etude preliminaire des lesions cellulaires de la poliomyelite en culture in vitro sur une souche de fibroblastes humains, *Ann. Inst. Pasteur Paris*, 85, 264, 1953.
20. **Barski, G., Robineaux, R., and Endo, M.**, Phase contrast cinematography of cellular lesion produced by poliomyelitis virus in vitro, *Proc. Soc. Exp. Biol. Med.*, 88, 57, 1955.
21. **Beale, A. J., Stevens, P. F., Davis, N., Stackiw, W., and Rhodes, A. J.**, The development of inclusions in tissue cultures of monkey kidney epithelial cells infected with poliomyelitis virus, *Can. J. Microbiol.*, 2, 298, 1956.
22. **Mayor, H. D.**, Cytochemical and fluorescent antibody studies on the growth of poliovirus in tissue culture, *Tex. Rep. Biol. Med.*, 19, 106, 1961.
23. **Ruska, H., Stuart, D. C., and Winsser, J.**, Electron microscopic visualization of intranuclear virus-like bodies in epithelial cells infected with poliomyelitis virus, *Arch. Gesamte Virusforsch.*, 6, 379, 1955.
24. **Anzai, T. and Ozaki, Y.**, Intranuclear crystal formation of poliovirus: electron microscopic observations, *Exp. Mol. Pathol.*, 10, 176, 1969.
25. **Kawanishi, M. and Ozaki, Y.**, Intranuclear crystals in FL cells infected with echovirus type 6, *Jpn. J. Microbiol.*, 17, 524, 1973.
26. **Kawanishi, M.**, Intranuclear crystal formation in picornavirus-infected cells, *Arch. Virol.*, 57, 123, 1978.
27. **Dales, S., Eggers, H. J., Tamm, I., and Palade, G. E.**, Electron microscopic study of the formation of poliovirus, *Virology*, 26, 379, 1965.
28. **Luria, S. E., Darnell, J. E., Baltimore, D., and Campbell, A.**, *General Virology*, 3rd ed., John Wiley & Sons, New York, 1978, 312.

29. **Pollack, R. and Goldman, R.**, Synthesis of infective poliovirus in BSC-1 monkey cells enucleated with cytochalasin B, *Science*, 179, 915, 1973.
30. **Follett, E. A., Pringle, C. R., and Pennington, T. H.**, Virus development in enucleate cells: echovirus, poliovirus, pseudorabies virus, reovirus, respiratory syncytial virus and Semliki Forest virus, *J. Gen. Virol.*, 26, 183, 1975.
31. **Mattern, C. F. and Daniel, W. A.**, Replication of poliovirus in HeLa cells: electron microscopic observations, *Virology*, 26, 646, 1965.
32. **Bablanian, R., Eggers, H. J., and Tamm, I.**, Studies on the mechanism of poliovirus-induced cell damage. I. The relation between poliovirus-induced metabolic and morphological alterations in cultured cells, *Virology*, 26, 100, 1965.
33. **Bablanian, R., Eggers, H. J., and Tamm, I.**, Studies on the mechanism of poliovirus-induced cell damage. II. The relation between poliovirus growth and virus-induced morphological changes in cells, *Virology*, 26, 114, 1965.
34. **Weiss, M. and Meyer, J.**, Comparison of the effects of coxsackievirus A9 and of actinomycin D on the nucleolar ultrastructure of monkey kidney cells, *J. Ultrastruct. Res.*, 38, 411, 1972.
35. **Malherbe, H.**, Vervet simian agent SA 2, *Lancet*, 2, 101, 1973.
36. **Bablanian, R.**, Structural and functional alterations in cultured cells infected with cytocidal viruses, *Prog. Med. Virol.*, 19, 40, 1975.
37. **Negri, A.**, Beitrag zum Studium der Aetiologie der Tollwuth, *Z. Hyg. Infectionskr.*, 43, 507, 1903.
38. **Meredith, C. D., Rossouw, A. P., and Koch, H. V.**, An unusual case of human rabies thought to be of chiropteran origin, *S. Afr. Med. J.*, 45, 767, 1971.
39. **Derakhshan, I., Bahmanyar, M., Noorsalehi, S., Fayaz, A., Mohammad, M., and Ahouraii, P.**, Light-microscopical diagnosis of rabies. A reappraisal, *Lancet*, 1, 302, 1978.
40. **Goodpasture, E. W.**, A study of rabies, with reference to a neural transmission of the virus in rabbits, and the structure and significance of Negri bodies, *Am. J. Pathol.*, 1, 547, 1925.
41. **Wiktor, T. J. and Clark, H. F.**, Growth of rabies virus in cell culture, in *The Natural History of Rabies*, Vol. 1, Baer, G. M., Ed., Academic Press, New York, 1975, chap. 9.
42. **Matsumoto, S.**, Electron microscope studies of rabies virus in mouse brain, *J. Cell Biol.*, 19, 565, 1963.
43. **Love, R., Fernandes, M. V., and Koprowski, H.**, Cytochemistry of inclusion bodies in tissue culture cells infected with rabies virus, *Proc. Soc. Exp. Biol. Med.*, 116, 560, 1964.
44. **Matsumoto, S., Schneider, L. G., Kawai, A., and Yonezawa, T.**, Further studies on the replication of rabies and rabies-like viruses in organized cultures of mammalian neural tissues, *J. Virol.*, 14, 981, 1974.
45. **Atanasiu, P.**, Animal inoculation and the Negri body, in *The Natural History of Rabies*, Vol. 1, Baer, G. M., Ed., Academic Press, New York, 1975, chap. 19.
46. **Miyamoto, K. and Matsumoto, S.**, The nature of the Negri body, *J. Cell Biol.*, 27, 677, 1965.
47. **Fernandes, M. V., Wiktor, T. J., and Koprowski, H.**, Mechanism of the cytopathic effect of rabies virus in tissue culture, *Virology*, 21, 128, 1963.
48. **Martini, G. A. and Siegert, R.**, Eds., *Marburg Virus Disease*, Springer-Verlag, Berlin, 1971.
49. **Siegert, R.**, Marburg virus, *Virol. Monogr.*, 11, 97, 1972.
50. **Simpson, D. I., Bowen, E. T., and Bright, W. F.**, Vervet monkey disease: experimental infection of monkeys with the causative agent, and antibody studies in wild-caught monkeys, *Lab. Anim.*, 2, 75, 1968.
51. **Strickland-Cholmley, M. and Malherbe, H.**, Examination of South African primates for the presence of Marburg virus, in *Marburg Virus Disease*, Martini, G. A. and Siegert, R., Eds., Springer-Verlag, Berlin, 1971, 195.
52. **Slenczka, W., Wolff, G., and Siegert, R.**, A critical study of monkey sera for the presence of antibody against the Marburg virus, *Am. J. Epidemiol.*, 93, 496, 1971.
53. **Gear, J. S., Cassel, G. A., Gear, A. J., Trappler, B., Clausen, L., Meyers, A. M., Kew, M. C., Bothwell, T. H., Sher, R., Miller, G. B., Schneider, J., Koornhof, H. J., Gomperts, E. D., Isaacson, M., and Gear, J. H.**, Outbreak of Marburg virus disease in Johannesburg, *Br. Med. J.*, 4, 489, 1975.
54. **Hennessen, W.**, A hemorrhagic disease transmitted from monkeys to man, *Nat. Cancer Inst. Monogr.*, 29, 161, 1968.
55. **Hofmann, H. and Kunz, C.**, Cultivation of the Marburg virus (*Rhabdovirus simiae*) in cell cultures, in *Marburg Virus Disease*, Martini, G. A. and Siegert, R., Eds., Springer-Verlag, Berlin, 1971, 112.
56. **Malherbe, H. and Strickland-Cholmley, M.**, Human disease from monkeys (Marburg virus), *Lancet*, 1, 1434, 1968.
57. **Malherbe, H. and Strickland-Cholmley, M.**, Studies on the Marburg virus, in *Marburg Virus Disease*, Martini, G. A. and Siegert, R., Eds., Springer-Verlag, Berlin, 1971, 188.
58. **Malherbe, H., Harwin, R., and Ulrich, M.**, The cytopathic effects of vervet monkey viruses, *S. Afr. Med. J.*, 37, 407, 1963.
59. Center for Disease Control, U.S.A., Hemorrhagic fever — Africa, *Morbid. Mortal. Wkly. Rep.*, 25, 324, 1976.

60. Center for Disease Control, U.S.A., Update on viral hemorrhagic fever — Africa, *Morbid. Mortal. Wkly. Rep.*, 25, 339, 1976.
61. **Johnson, K. M., Webb, P. A., Lange, J. V., and Murphy, F. A.**, Isolation and partial characterisation of a new virus causing acute haemorrhagic fever in Zaire, *Lancet*, 1, 569, 1977.
62. **Bowen, E. T., Platt, G. S., Lloyd, G., Baskerville, A., Harris, W. J., and Vella, E. E.**, Viral haemorrhagic fever in southern Sudan and northern Zaire, *Lancet*, 1, 571, 1977.
63. **Pattyn, S., Jacob, W., van der Groen, G., Piot, P., and Courteille, G.**, Isolation of Marburg-like virus from a case of haemorrhagic fever in Zaire, *Lancet*, 1, 573, 1977.
64. **Sabin, A. B.**, Reoviruses, *Science*, 130, 1387, 1959.
65. **Rosen, L.**, Reoviruses, *Virol. Monogr.*, 1, 73, 1968.
66. **Rhim, J. S., Smith, K. O., and Melnick, J. L.**, Complete and coreless forms of reovirus (ECHO 10), *Virology*, 15, 428, 1961.
67. **Lerner, A. M., Cherry, J. D., and Finland, M.**, Enhancement of cytopathic effects of reoviruses in rolled cultures of rhesus kidney, *Proc. Soc. Exp. Biol. Med.*, 110, 727, 1962.
68. **Malherbe, H. and Harwin, R.**, Seven viruses isolated from the vervet monkey, *Br. J. Exp. Pathol.*, 38, 539, 1957.
69. **Love, R. and Fernandes, M.**, Cytochemistry of virus-induced inclusions containing nucleic acids, *Proc. Soc. Exp. Biol. Med.*, 136, 1072, 1971.
70. **Tournier, P. and Plissier, M.**, Le developpement intracellulaire du reovirus observe au microscope electronique, *Presse Med.*, 68, 683, 1960.
71. **Rhim, J. S., Jordan, L. E., and Mayor, H. D.**, Cytochemical, fluorescent-antibody and electron microscopic studies on the growth of reovirus (ECHO 10) in tissue culture, *Virology*, 17, 342, 1962.
72. **Malherbe, H. H. and Strickland-Cholmley, M.**, Quantitative studies on viral survival in sewage purification processes, in *Transmission of Viruses by the Water Route*, Berg, G., Ed., Interscience, New York, 1967, 379.
73. **Malherbe, H. H., Strickland-Cholmley, M., and Geyer, S. M.**, Viruses in abattoir effluents, in *Transmission of Viruses by the Water Route*, Berg, G., Ed., Interscience, New York, 1967, 347.
74. **Flewett, T. H., Bryden, A. S., Davies, H., Woode, G. N., Bridger, J. C., and Derrick, J. M.**, Relation between viruses from acute gastroenteritis of children and newborn calves, *Lancet*, 2, 61, 1974.
75. **Bishop, R. F., Davidson, G. P., Holmes, I. H., and Ruck, B. J.**, Virus particles in epithelial cells of duodenal mucosa from children with acute non-bacterial gastroenteritis, *Lancet*, 2, 1281, 1973.
76. **Flewett, T. H., Davies, H., Bryden, A. S., and Robertson, M. J.**, Diagnostic electron microscopy of faeces. II. Acute gastroenteritis associated with reovirus-like particles, *J. Clin. Pathol.*, 27, 608, 1974.
77. **Fernelius, A. L., Ritchie, A. E., Classick, L. G., Norman, J. O., and Mebus, C. A.**, Cell culture adaptation and propagation of a reovirus-like agent of calf diarrhea from a field outbreak in Nebraska, *Arch. Gesamte Virusforsch.*, 37, 114, 1972.
78. **Cheever, F. S. and Mueller, J. H.**, Epidemic diarrheal disease of suckling mice. I. Manifestations, epidemiology, and attempts to transmit the disease, *J. Exp. Med.*, 85, 405, 1947.
79. **Pappenheimer, A. M. and Enders, J. F.**, An epidemic diarrheal disease of suckling mice. II. Inclusions in the intestinal epithelial cells, *J. Exp. Med.*, 85, 417, 1947.
80. **Pappenheimer, A. M. and Cheever, F. S.**, Epidemic diarrheal disease of suckling mice. IV. Cytoplasmic inclusion bodies in intestinal epithelium in relation to the disease, *J. Exp. Med.*, 88, 317, 1948.
81. **Kraft, L. M.**, Studies on the etiology and transmission of epidemic diarrhea of infant mice, *J. Exp. Med.*, 106, 743, 1957.
82. **Kraft, L. M.**, Observations on the control and natural history of epidemic diarrhea of infant mice (EDIM), *Yale J. Biol. Med.*, 31, 121, 1958.
83. **Adams, W. R. and Kraft, L. M.**, Epizootic diarrhea of infant mice: identification of the etiologic agent, *Science*, 141, 359, 1963.
84. **Banfield, W. G., Kasnic, G., and Blackwell, J. H.**, Further observations on the virus of epizootic diarrhea of infant mice. An electron microscopic study, *Virology*, 36, 411, 1968.
85. **Rubenstein, D., Milne, R. G., Buckland, R., and Tyrrell, D. A.**, The growth of the virus of epidemic diarrhoea of infant mice (EDIM) in organ-cultures of intestinal epithelium, *Br. J. Exp. Pathol.*, 52, 442, 1971.
86. **Woode, G. N., Bridger, J. C., Jones, J. M., Flewett, T. H., Bryden, A. S., Davies, H. A., and White, G. B.**, Morphological and antigenic relationships between viruses (Rotaviruses) from acute gastroenteritis of children, calves, piglets, mice, and foals, *Infect. Immun.*, 14, 804, 1976.
87. **Malherbe, H. H. and Strickland-Cholmley, M.**, Simian virus SA 11 and the related O agent, *Arch. Gesamte Virusforsch.*, 22, 235, 1967.
88. **Els, H. J. and Lecatsas, G.**, Morphological studies on Simian virus SA 11 and the "related" O Agent, *J. Gen. Virol.*, 17, 129, 1972.

89. **Lecatsas, G.**, Virus particles in acute gastroenteritis, *Lancet*, 2, 524, 1974.
90. **Kapikian, A. Z., Cline, W. L., Kim, H. W., Kalica, A. R., Wyatt, R. G., Vankirk, D. H., Chanock, R. M., James, H. D., and Vaughn, A. L.**, Antigenic relationships among five reovirus-like (RVL) agents by complement fixation (CF) and development of new substitute CF antigens for the human RVL agent of infantile gastroenteritis, *Proc. Soc. Exp. Biol. Med.*, 152, 535, 1976.
91. **Mebus, C. A., Kono, M., Underdahl, N. R., and Twiehaus, M. J.**, Cell culture propagation of neonatal calf diarrhea (scours) virus, *Can. Vet. J.*, 12, 69, 1971.
92. **Welch, A. B. and Twiehaus, M. J.**, Cell culture studies of a neonatal calf diarrhea virus, *Can. J. Comp. Med.*, 37, 287, 1973.
93. **Ozawa, Y. and Hazrati, A.**, Growth of African horse-sickness virus in monkey kidney cell cultures, *Am. J. Vet. Res.*, 25, 505, 1964.
94. **Ozawa, Y., Hopkins, I. G., Hazrati, A., Modjitabai, A., and Kaveh, P.**, Cytology of monkey kidney cells infected with African horse-sickness virus, *Nature (London)*, 206, 1321, 1965.
95. **Ozawa, Y., Mojtabai, A., Hopkins, I. G., Hazrati, A., and Kaveh, P.**, Sequential cellular changes produced by African horse-sickness virus in monkey kidney cells, *Am. J. Vet. Res.*, 27, 558, 1966.
96. **Ozawa, Y.**, Studies on the replication of African horse-sickness virus in two different cell line cultures, *Arch. Gesamte Virusforsch.*, 21, 155, 1967.
97. **Breese, S. S. and Ozawa, Y.**, Intracellular inclusions resulting from infection with African horsesickness virus, *J. Virol.*, 4, 109, 1969.
98. **Lecatsas, G. and Erasmus, B. J.**, Electron microscopic study of the formation of African Horse-sickness virus, *Arch. Gesamte Virusforsch.*, 22, 442, 1967.
99. **Haig, D. A., McKercher, D. G., and Alexander, R. A.**, The cytopathogenic action of bluetongue virus on tissue cultures and its application to the detection of antibodies in the serum of sheep, *Onderstepoort J. Vet. Res.*, 27, 171, 1956.
100. **Fernandes, M. V.**, Isolation and propagation of bluetongue virus in tissue culture, *Am. J. Vet. Res.*, 20, 398, 1959.
101. **Livingston, C. W. and Moore, R. W.**, Cytochemical changes of bluetongue virus in tissue cultures, *Am. J. Vet. Res.*, 23, 701, 1962.
102. **Bowne, J. G. and Jochim, M. M.**, Cytopathologic changes and development of inclusion bodies in cultured cells infected with bluetongue virus, *Am. J. Vet. Res.*, 28, 1091, 1967.
103. **Bowne, J. G.**, Bluetongue disease, *Adv. Vet. Sci. Comp. Med.*, 15, 1, 1971.
104. **Howell, P. G. and Verwoerd, D. W.**, Bluetongue virus, *Virol. Monogr.*, 9, 35, 1971.
105. **Ichihashi, Y. and Matsumoto, S.**, The relationship between poxvirus and A-type inclusion body during double infection, *Virology*, 36, 262, 1968.
106. **Kato, S., Hara, J., Ogawa, M., Miyamoto, H., and Kamahora, J.**, Inclusion markers of cowpox virus and alastrim virus, *Biken J.*, 6, 233, 1963.
107. **Kato, S., Takahashi, M., Kameyama, S., and Kamahora, J.**, A study on the morphological and cyto-immunological relationship between the inclusions of variola, cowpox, rabbitpox, vaccinia (variola origin) and vaccinia IHD and a consideration of the term "Guarnieri body", *Biken J.*, 2, 353, 1959.
108. **Kato, S. and Cutting, W.**, A study of the inclusion bodies of rabbit myxoma and fibroma virus and a consideration of the relationship between all pox virus inclusion bodies, *Stanford Med. Bull.*, 17, 34, 1959.
109. **Barski, G.**, Detection of latent ectromelia virus infection in mice with the aid of human cell cultures in vitro, *Virology*, 10, 155, 1960.
110. **Kaku, H. and Kamahora, J.**, Giant cell formation in L cells infected with active vaccinia virus, *Biken J.*, 6, 299, 1964.
111. **Ichihashi, Y., Matsumoto, S. and Dales, S.**, Biogenesis of poxviruses: role of A-type inclusions and host cell membranes in virus dissemination, *Virology*, 46, 507, 1971.
112. **Tsuchiya, Y. and Tagaya, I.**, Isolation of a multinucleated giant cell-forming and hemagglutinin-negative variant of variola virus, *Arch. Gesamte Virusforsch.*, 39, 292, 1972.
113. **Baxby, D.**, Differentiation of smallpox and camelpox viruses in cultures of human and monkey cells, *J. Hyg.*, 72, 251, 1974.
114. **Baxby, D.**, Identification and interrelationships of the variola/vaccinia subgroup of poxviruses, *Prog. Med. Virol.*, 19, 215, 1975.
115. **Hahon, N.**, Cytopathogenicity and propagation of variola virus in tissue culture, *J. Immunol.*, 81, 426, 1958.
116. **Baxby, D.**, The origins of vaccinia virus, *J. Infect. Dis.*, 136, 453, 1977.
117. **Hruby, D. E., Guarino, L. A., and Kates, J. R.**, Vaccinia virus replication. I. Requirement for the host-cell nucleus, *J. Virol.*, 29, 705, 1979.
118. **Marchal, J.**, Infectious ectromelia. A hitherto undescribed virus disease of mice, *J. Pathol. Bacteriol.*, 33, 713, 1930.
119. **Downie, A. W. and McGaughey, C. A.**, The cultivation of the virus of infectious ectromelia, with observations on the formation of inclusion bodies in vitro, *J. Pathol. Bacteriol.*, 40, 147, 1935.

120. **Ichihashi, Y. and Matsumoto, S.**, Studies on the nature of Marchal bodies (A-type inclusion) during ectromelia virus infection, *Virology*, 29, 264, 1966.
121. **Kamahora, J., Toyoshima, K., Nii, S., and Kato, S.**, A study on the number of infectious virus particles necessary for "B" type inclusion formation, *Biken J.*, 2, 345, 1959.
122. **Nii, S.**, Plaque formation and some cytopathogenic characteristics of ectromelia virus (Hampstead strain) and vaccinia virus (IHD strain) on monolayers, *Biken J.*, 2, 195, 1959.
123. **Rouhandeh, H., Engler, R., Fouad, M. T., and Sells, L. L.**, Properties of monkey pox virus, *Arch. Gesamte Virusforsch.*, 20, 363, 1967.
124. **Cho, C. T. and Wenner, H. A.**, In vitro growth characteristics of monkeypox virus, *Proc. Soc. Exp. Biol. Med.*, 139, 916, 1972.
125. **Cho, C. T. and Wenner, H. A.**, Monkeypox virus, *Bacteriol. Rev.*, 37, 1, 1973.
126. **Ladnyj, I. D., Ziegler, P., and Kima, E.**, A human infection caused by monkeypox virus in Basankusu Territory, Democratic Republic of the Congo, *Bull. W.H.O.*, 46, 593, 1972.
127. World Health Organization, Monkeypox virus infections, *W.H.O. Wkly. Epidem. Rec.*, 19, 149, 1976.
128. Editorial, Smallpox research after Birmingham, *Lancet*, 2, 560, 1978.
129. Center for Disease Control, U.S.A., Monkeypox in humans — West Africa, *Morbid. Mortal. Wkly. Rep.*, 28, 135, 1979.
130. **Arita, I. and Henderson, D. A.**, Smallpox and monkeypox in non-human primates, *Bull. W.H.O.*, 39, 277, 1968.
131. **Bourke, A. T. and Dumbell, K. R.**, An unusual poxvirus from Nigeria, *Bull. W.H.O.*, 46, 621, 1972.
132. **Lourie, B., Nakano, J. H., Kemp, G. E., and Setzer, H. W.**, Isolation of poxvirus from an African rodent, *J. Infect. Dis.*, 132, 677, 1975.
133. **Bearcroft, W. G. and Jamieson, M. F.**, An outbreak of subcutaneous tumours in rhesus monkeys, *Nature (London)*, 182, 195, 1958.
134. **Andrewes, C. H., Allison, A. C., Armstrong, J. A., Bearcroft, G., Niven, J. S., and Pereira, H. G.**, A virus disease of monkeys causing large superficial growths, *Acta Unio Int. Contra Cancrum*, 15, 760, 1959.
135. **Yohn, D. S., Grace, J. T., and Haendiges, V. A.**, Immunofluorescent and histochemical study of Yaba tumor virus replication in cell culture, *Bacteriol. Proc.*, 132, 1964.
136. **Yohn, D. S., Grace, J. T., and Haendiges, V. A.**, A quantitative cell culture assay for Yaba tumour virus, *Nature (London)*, 202, 881, 1964.
137. **Levinthal, J. M. and Shein, H. M.**, Propagation of a simian tumor agent (Yaba virus) in cultures of human and simian renal cells as detected by immunofluorescence, *Virology*, 23, 268, 1964.
138. **Downie, A. W., Taylor-Robinson, C. H., Caunt, A. E., Nelson, G. S., Manson-Bahr, P. E., and Matthews, T. C.**, Tanapox: a new disease caused by a pox virus, *Br. Med. J.*, 1, 363, 1971.
139. **Hall, A. S. and McNulty, W. P.**, A contagious pox disease of monkeys, *J. Am. Vet. Med. Assoc.*, 151, 833, 1967.
140. **Downie, A. W. and Espana, C.**, Comparison of Tanapox virus and Yaba-like viruses causing epidemic disease in monkeys, *J. Hyg.*, 70, 23, 1972.
141. **Tsuchiya, Y. and Tagaya, I.**, Sero-epidemiological survey on Yaba and 1211 virus infections among several species of monkeys, *J. Hyg.*, 69, 445, 1971.
142. **Downie, A. W.**, Serological evidence of infection with Tana and Yaba poxviruses among several species of monkey, *J. Hyg.*, 72, 245, 1974.
143. **Casey, H. W., Woodruff, J. M., and Butcher, W. I.**, Electron microscopy of a benign epidermal pox disease of rhesus monkeys, *Am. J. Pathol.*, 51, 431, 1967.
144. **Nicholas, A. H. and McNulty, W. P.**, In vitro characteristics of a poxvirus isolated from rhesus monkeys, *Nature (London)*, 217, 745, 1968.
145. **Nicholas, A. H.**, A poxvirus of primates. I. Growth of the virus in vitro and comparison with other poxviruses, *J. Nat. Cancer Inst.*, 45, 897, 1970.
146. **de Boer, G. F.**, Swinepox. Virus isolation, experimental infections and the differentiation from vaccinia virus infections, *Arch. Virol.*, 49, 141, 1975.
147. **Teppema, J. S. and de Boer, G. F.**, Ultrastructural aspects of experimental swinepox with special reference to inclusion bodies, *Arch. Virol.*, 49, 151, 1975.
148. **Wildy, P.**, Classification and nomenclature of viruses, *Monogr. Virol.*, 5, 73, 1971.
149. **Rowe, W. P., Murphy, F. A., Bergold, G. H., Casals, J., Hotchin, J., Johnson, K. M., Lehmann-Grube, F., Mims, C. A., Traub, E., and Webb, P. A.**, Arenoviruses: proposed name for a newly defined virus group, *J. Virol.*, 5, 651, 1970.
150. **Lehmann-Grube, F.**, Lymphocytic choriomeningitis virus, *Virol. Monogr.*, 10, 1, 1971.
151. **Benda, R. and Cinatl, J.**, Multiplication of Lymphocytic choriomeningitis virus in bottle cell cultures. Experimental data for the preparation of highly infectious fluids, *Acta Virol. Engl. Ed.*, 5, 159, 1962.
152. **Casals, J.**, Arenaviruses, *Yale J. Biol. Med.*, 48, 115, 1975.

153. **Mettler, N., Buckley, S. M., and Casals, J.**, Propagation of Junin virus, the etiological agent of Argentinian hemorrhagic fever, in HeLa cell cultures, *Proc. Soc. Exp. Biol. Med.*, 107, 684, 1961.
154. **Buckley, S. M.**, Junin and Tacaribe work in HeLa cells, *Am. J. Trop. Med. Hyg.*, 14, 792, 1965.
155. **Murphy, F. A., Webb, P. A., Johnson, K. M., Whitfield, S. G., and Chappell, W. A.**, Arenoviruses in Vero cells: ultrastructural studies, *J. Virol.*, 6, 507, 1970.
156. **Buckley, S. M., Casals, J., and Downs, W. G.**, Isolation and antigenic characterization of Lassa virus, *Nature (London)*, 227, 174, 1970.
157. **Speir, R. W., Wood, O., Liebhaber, H., and Buckley, S. M.**, Lassa fever, a new virus disease of man from West Africa. IV. Electron microscopy of Vero cell cultures infected with Lassa virus, *Am. J. Trop. Med. Hyg.*, 19, 692, 1970.
158. **Fenner, F.**, The classification and nomenclature of viruses, *J. Gen. Virol.*, 31, 463, 1976.
159. **Dalton, A. J., Melnick, J. L., Bauer, H., Beaudreau, G., Bentvelzen, P., Bolognesi, D., Gallo, R., Graffi, A., Haguenau, F., Heston, W., Huebner, R., Todaro, G., and Heine, U. I.**, The case for a family of reverse transcriptase viruses: retraviridae, *Intervirology*, 4, 201, 1974.
160. **Enders, J. F. and Peebles, T. C.**, Propagation in tissue cultures of cytopathogenic agents from patients with measles, *Proc. Soc. Exp. Biol. Med.*, 86, 277, 1954.
161. **Rustigian, R., Johnston, P., and Reihart, H.**, Infection of monkey kidney tissue cultures with virus-like agents, *Proc. Soc. Exp. Biol. Med.*, 88, 8, 1955.
162. **Johnston, P. B.**, Taxonomic features of seven serotypes of simian and ape foamy viruses, *Infect. Immun.*, 3, 793, 1971.
163. **Johnston, P. B.**, Studies on simian foamy viruses and syncytium-forming viruses of lower animals, *Lab. Anim. Sci.*, 24, 159, 1974.
164. **Hooks, J. J. and Gibbs, C. J.**, The foamy viruses, *Bacteriol. Rev.*, 39, 169, 1975.
165. **Hsiung, G. D.**, Latent virus infections in primate tissues with special reference to simian viruses, *Bacteriol. Rev.*, 32, 185, 1968.
166. **Achong, B. G., Mansell, P. W., Epstein, M. A., and Clifford, P.**, An unusual virus from a human nasopharyngeal carcinoma, *J. Nat. Cancer Inst.*, 46, 299, 1971.
167. **Epstein, M. A., Achong, B. G., and Ball, G.**, Further observations on a human syncytial virus from a nasopharyngeal carcinoma, *J. Nat. Cancer Inst.*, 53, 681, 1974.
168. **Nemo, G. J., Brown, P. W., Gibbs, C. J., and Gajdusek, D. C.**, Antigenic relationship of human foamy virus to the simian foamy viruses, *Infect. Immun.*, 20, 69, 1978.
169. **Brown, P., Nemo, G., and Gajdusek, D. C.**, Human foamy virus: further characterization, seroepidemiology, and relationship to chimpanzee foamy viruses, *J. Infect. Dis.*, 137, 421, 1978.
170. **Young, D., Samuels, J., and Clarke, J. K.**, A foamy virus of possible human origin isolated in BHK-21 cells, *Arch. Gesamte Virusforsch.*, 42, 228, 1973.
171. **Cameron, K. R., Birchall, S. M., and Moses, M. A.**, Isolation of foamy virus from patient with dialysis encephalopathy, *Lancet*, 2, 796, 1978.
172. **Thormar, H., Lin, F. H., and Trowbridge, R. S.**, Visna and Maedi viruses in tissue culture, *Prog. Med. Virol.*, 18, 323, 1974.
173. **Haase, A. T.**, The slow infection caused by Visna virus, *Curr. Top. Microbiol. Immunol.*, 72, 101, 1975.
174. **Gudnadottir, M.**, Visna-Maedi in sheep, *Prog. Med. Virol.*, 18, 336, 1974.
175. **Thormar, H. and Helgadottir, H.**, A comparison of Visna and Maedi viruses. II. Serological relationship, *Res. Vet. Sci.*, 6, 456, 1965.
176. **Sigurdsson, B., Thormar, H., and Palsson, P. A.**, Cultivation of Visna virus in tissue culture, *Arch. Gesamte Virusforsch.*, 10, 368, 1960.
177. **Sigurdardottir, B. and Thormar, H.**, Isolation of a viral agent from the lungs of sheep affected with Maedi, *J. Infect. Dis.*, 114, 55, 1964.
178. **Thormar, H. and Sigurdardottir, B.**, Growth of Visna virus in primary tissue cultures from various animal species, *Acta Pathol. Microbiol. Scand.*, 55, 180, 1962.
179. **Haase, A. T., Stowring, L., Narayan, O., Griffin, D., and Price, D.**, Slow persistent infection caused by Visna virus: role of host restriction, *Science*, 195, 175, 1977.
180. **Macintyre, E. H., Wintersgill, C. J., and Thormar, H.**, Morphological transformation of human astrocytes by Visna virus with complete virus production, *Nature (London) New Biol.*, 237, 111, 1972.
181. **August, M. J. and Harter, D. H.**, Visna virus-induced fusion of continuous simian kidney cells, *Arch. Gesamte Virusforsch.*, 44, 92, 1974.
182. **Melnick, J. L.**, The classification of viruses, in *Human Diseases Caused by Viruses*, Rothschild, H., Allison, F., and Howe, C., Eds., Oxford University Press, New York, 1978, chap. 2.
183. **Klement, V., Rowe, W. P., Hartley, J. W., and Pugh, W. E.**, Mixed culture cytopathogenicity: a new test for growth of murine leukemia viruses in tissue culture, *Proc. Nat. Acad. Sci. U.S.A.*, 63, 753, 1969.

184. **Rangan, S. R., Ueberhorst, P. J., and Wong, M. C.,** Syncytial giant cell focus assay for viruses derived from feline leukemia and a simian sarcoma, *Proc. Soc. Exp. Biol. Med.*, 142, 1077, 1973.
185. **Klement, V. and McAllister, R. M.,** Syncytial cytopathic effect in KB cells of a C-type RNA virus isolated from human rhabdomyosarcoma, *Virology*, 50, 305, 1972.
186. **Hampar, B., Rand, K. H., Lerner, R. A., Del Villano, B. C., McAllister, R. M., Martos, L. M., Derge, J. G., Long, C. W., and Gilden, R. V.,** Formation of syncytia in human lymphoblastoid cells infected with Type C viruses, *Virology*, 55, 453, 1973.
187. **Rand, K. H., Reed, B., Broseus, I., Hampar, B., Gilden, R., and Long, C.,** Mixed-culture cytopathogenicity between KC and XC oncornavirus indicator lines and "virus-free" human choriocarcinoma cells, *J. Nat. Cancer Inst.*, 56, 509, 1976.
188. **Ogura, H., Tanaka, T., Ocho, M., Kuwata, T., and Oda, T.,** Detection of Mason-Pfizer monkey virus infection by syncytia formation of human cells doubly transformed by Rous sarcoma virus and simian virus 40, *Arch. Virol.*, 57, 195, 1978.
189. **Zhdanov, V. M., Soloviev, V. D., Bektemirov, T. A., Ilyin, K. V., Bykovsky, A. F., Mazurenko, N. P., Irlin, I. S., and Yershov, F. I.,** Isolation of oncornaviruses from continuous human cell cultures, *Intervirology*, 1, 19, 1973.
190. **Lieber, M. M., Benveniste, R. E., Livingston, D. M., and Todaro, G. J.,** Mammalian cells in culture frequently release Type C viruses, *Science*, 182, 56, 1973.
191. **Weiss, R. A.,** Why cell biologists should be aware of genetically transmitted viruses, *Nat. Cancer Inst. Monogr.*, 48, 183, 1978.
192. **Bradburne, A. F. and Tyrrell, D. A. J.,** The propagation of coronaviruses in tissue-culture, *Arch. Gesamte. Virusforsch.*, 28, 133, 1969.
193. **McIntosh, K.,** Coronaviruses: a comparative review, *Curr. Top. Microbiol. Immunol.*, 63, 85, 1974.
194. **Morris, J. A., Blount, R. E., and Savage, R. E.,** Recovery of cytopathogenic agent from chimpanzees with coryza, *Proc. Soc. Exp. Biol. Med.*, 92, 544, 1956.
195. **Chanock, R., Roizman, B., and Myers, R.,** Recovery from infants with respiratory illness of a virus related to chimpanzee coryza agent (CCA). I. Isolation, properties, and characterization, *Am. J. Hyg.*, 66, 281, 1957.
196. **Chanock, R. and Finberg, L.,** Recovery from infants with respiratory illness of a virus related to chimpanzee coryza agent (CCA). II. Epidemiologic aspects of infection in infants and young children, *Am. J. Hyg.*, 66, 291, 1957.
197. **Jordan, W. S.,** Growth characteristics of Respiratory Syncytial virus, *J. Immunol.*, 88, 581, 1962.
198. **Armstrong, J. A., Pereira, H. G., and Valentine, R. C.,** Morphology and development of Respiratory Syncytial virus in cell cultures, *Nature (London)*, 196, 1179, 1962.
199. **Norrby, E., Marusyk, H., and Orvell, C.,** Morphogenesis of Respiratory Syncytial virus in a green monkey kidney cell line (Vero), *J. Virol.*, 6, 237, 1970.
200. **Corbitt, G. and Beswick, T. S.,** The cytochemistry of inclusion bodies in cell cultures into which Respiratory Syncytial virus has been inoculated, *J. Med. Microbiol.*, 5, 144, 1972.
201. **Gharpure, M. A., Wright, P. F., and Chanock, R. M.,** Temperature-sensitive mutants of Respiratory Syncytial virus, *J. Virol.*, 3, 414, 1969.
202. **Belshe, R. B., Richardson, L. S., Schnitzer, T. J., Prevar, D. A., Camargo, E., and Chanock, R. M.,** Further characterization of the complementation Group B temperature-sensitive mutant of Respiratory Syncytial virus, *J. Virol.*, 24, 8, 1977.
203. **Poste, G. and Allison, A. C.,** Membrane fusion, *Biochim. Biophys. Acta*, 300, 421, 1973.
204. **Poste, G.,** Virus-induced polykaryocytosis and the mechanism of cell fusion, *Adv. Virus Res.*, 16, 303, 1970.
205. **Poste, G.,** Mechanisms of virus-induced cell fusion, *Int. Rev. Cytol.*, 33, 157, 1972.
206. **Bratt, M. A. and Gallaher, W. R.,** Preliminary analysis of the requirements for fusion from within and fusion from without by Newcastle disease virus, *Proc. Nat. Acad. Sci. U.S.A.*, 64, 536, 1969.
207. **Kingsbury, D. W., Bratt, M. A., Choppin, P. W., Hanson, R. P., Hosaka, Y., Ter Meulen, V., Norrby, E., Plowright, W., Rott, R., and Wunner, W. H.,** Paramyxoviridae, *Intervirology*, 10, 137, 1978.
208. **Brandt, C. D.,** Inclusion body formation with Newcastle disease and mumps viruses in cultures of chick embryo cells, *Virology*, 5, 177, 1958.
209. **Brandt, C. D.,** Cytopathic action of myxoviruses on cultivated mammalian cells, *Virology*, 14, 1, 1961.
210. **Poste, G., Alexander, D. J., Reeve, P., and Hewlett, G.,** Modification of Newcastle disease virus release and cytopathogenicity in cells treated with plant lectins, *J. Gen. Virol.*, 23, 255, 1974.
211. **Gresser, I. and Enders, J. F.,** Cytopathogenicity of mumps virus in cultures of chick embryo and human amnion cells, *Proc. Soc. Exp. Biol. Med.*, 107, 804, 1961.
212. **Nasibov, M. N. and Smorodintsev, A. A.,** An investigation into the nature of the intracellular inclusions formed by mumps virus, as determined by the combined use of immunofluorescence and methylgreen-pyronin stain, *Arch. Gesamte Virusforsch.*, 24, 319, 1968.

213. **Watson, B. K.,** Distribution of mumps virus in tissue cultures as determined by fluorescein-labeled antiserum, *Proc. Soc. Exp. Biol. Med.*, 79, 222, 1952.
214. **Traver, M. I., Northrop, R. L., and Walker, D. L.,** Site of intracellular antigen production by myxoviruses, *Proc. Soc. Exp. Biol. Med.*, 104, 268, 1960.
215. **Frank, A. L., Couch, R. B., Griffis, C. A., and Baxter, B. D.,** Comparison of different tissue cultures for isolation and quantitation of influenza and parainfluenza viruses, *J. Clin. Microbiol.*, 10, 32, 1979.
216. **Silver, S. M., Scheid, A., and Choppin, P. W.,** Loss on serial passage of rhesus monkey kidney cells of proteolytic activity required for Sendai virus activity, *Infect. Immun.*, 20, 235, 1978.
217. **Smorodintsev, A. A.,** Experiences with the isolation and propagation of parainfluenza viruses, *Acta Virol.*, 6, 338, 1962.
218. **Canchola, J., Vargosko, A. J., Kim, H. W., Parrott, R. H., Christmas, E., Jeffries, B., and Chanock, R. M.,** Antigenic variation among newly isolated strains of parainfluenza Type 4 virus, *Am. J. Hyg.*, 79, 357, 1964.
219. **Hsiung, G. D.,** Parainfluenza-5 virus. Infection of man and animal, *Prog. Med. Virol.*, 14, 241, 1972.
220. **Chanock, R. M., Johnson, K. M., Cook, M. K., Wong, D. C., and Vargosko, A.,** The hemadsorption technique, with special reference to the problem of naturally occurring simian parainfluenza virus, *Am. Rev. Respir. Dis.*, 83, 125, 1961.
221. **Hull, R. N.,** The simian viruses, *Virol. Monogr.*, 2, 1, 1968.
222. **Choppin, P. W.,** Multiplication of a myxovirus (SV 5) with minimal cytopathic effects and without interference, *Virology*, 23, 224, 1964.
223. **Compans, R. W., Holmes, K. V., Dales, S., and Choppin, P. W.,** An electron microscopic study of moderate and virulent virus-cell interactions of the parainfluenza virus SV 5, *Virology*, 30, 411, 1966.
224. **Holmes, K. V. and Choppin, P. W.,** On the role of the response of the cell membrane in determining virus virulence. Contrasting effects of the parainfluenza virus SV 5 in two cell types, *J. Exp. Med.*, 124, 501, 1966.

Chapter 6

CYTOPATHIC CHANGES AFFECTING MAINLY THE NUCLEUS

I. ARBOVIRUSES

Many papers have been published on the growth of arboviruses in cell cultures of various kinds, but usually the cytopathic effects noted have been of a grossly overt nature only, comprising granularity and rounding of cells starting focally and then proceeding to complete destruction of the cell sheet. Plaques of degenerated cells under agar overlay have been used for arbovirus assays.

In 1930 Cowdry and Kitchen[1] drew attention to earlier reports of nuclear inclusions in liver cells infected with Yellow Fever virus. They examined histological sections of liver tissue taken from 50 monkeys infected with the virus, and listed the tinctorial reactions of the nuclear inclusions using a variety of stains. The cultivation in vitro of Yellow Fever virus has been reported by a number of workers,[2-5] but we have no information regarding the production in vitro of nuclear inclusions such as those noted by Cowdry and Kitchen in their histological sections.

Buckley and Karabatsos[5-7] tested 243 arboviruses in cell cultures, and of these only 14 failed to produce an overt cytopathic effect. Mussgay et al.[8] in their review of the growth cycles of arboviruses in vertebrate and arthropod cells have mentioned that St. Louis encephalitis and Louping ill viruses grow silently in some types of cell but produce overt changes in others. Syncytium formation by Japanese encephalitis virus[9] and by Eastern and Venezuelan equine encephalitis viruses[10] has been observed.

The Kemerovo group of tickborne orbiviruses has been reported[11,12] to produce cytoplasmic inclusions associated with viral assembly, which is a change characteristic of reoviridae, as noted in Chapter 5,III.C. Another tickborne virus from Nigeria was also found to form cytoplasmic inclusions in several mammalian cell lines.[13]

An exhaustive search of the literature would undoubtedly yield examples of other cytopathic changes produced by arboviruses, but in the main there appears to be relatively little information on the use of stained cultures. We have selected two instances of nuclear changes produced by arboviruses because these appear to be distinctive for arboviruses and serve also as examples of the range of cellular changes that can be observed in stained cultures.

A. Sindbis Virus

Sindbis virus was first isolated by Taylor et al.[14] at the Egyptian village of that name. In 1963 we reported[15] the isolation of Sindbis virus from skin lesions of a human patient, which was the first association of this virus with a distinctive clinical illness. The presence of small nuclear inclusions, in primary monkey kidney cell cultures we had inoculated with material from the skin lesions, indicated a probable virus infection. Subsequent inoculation of newborn mice with the production of lesions in connective tissue, which differentiated the agent from the coxsackieviruses, led us to review histological sections of arbovirus-infected mice. This narrowed the field to a few agents including Sindbis virus. Serological tests then confirmed the identity of the agent and established its etiological role in the clinical disease. Furthermore, the pathological lesions we observed in the muscles and connective tissue of newborn mice provided an exact parallel for the pathology underlying a number of the human signs and symptoms.

Viral replication takes place entirely in the cytoplasm,[16,17] but small eosinophilic inclusions occur in the nucleus. Infected cells undergo a nonspecific type of degeneration, but the multiple typical small inclusions indicate a virus infection. In Figure 46* we show the inclusions produced by Sindbis virus in primary vervet monkey kidney cells.

B. Rift Valley Fever Virus

This virus, a member of the arbovirus family of Bunyaviridae, is an important cause of disease in sheep and cattle in Africa and can also infect man. In 1963 Coackley[18] described the cytopathic changes caused by the virus in lamb testis cell cultures, and cited an earlier study by Plowright and Ferris in which mention was made of elongated intranuclear inclusions in sheep kidney cells infected with Rift Valley Fever virus. Coackley described two types of nuclear inclusion: one kind comprised fine eosinophilic fusiform filaments which occasionally appeared to fuse or branch; and the second kind consisted of small rounded eosinophilic bodies surrounded by basophilic chromatin-like material.

We have confirmed the production of eosinophilic filamentous nuclear inclusions in primary vervet monkey kidney cells infected with Rift Valley Fever virus, and we illustrate those in Figure 47. The darker round masses are nucleoli. In Figure 48 we show annular forms of the inclusions around the nucleoli. Both the filamentous and the annular forms are highly characteristic of Rift Valley Fever virus, and we have not observed similar inclusions in cells infected by any other virus.

II. PARVOVIRUSES

The family of Parvoviridae[19,20] includes the smallest known DNA viruses and has three genera: *Parvovirus*, which multiplies without the aid of an adenovirus; Adeno-associated virus which requires an adenovirus as helper for replication; and the *Densovirus* of insects. The adeno-associated agents do not produce a visible cytopathic effect, and we shall therefore consider here only the parvoviruses.

Several animal species harbor parvoviruses, including feline, porcine, bovine, and avian agents. These viruses have a number of features in common: there is a preference for growth in actively dividing cells, the nucleus is the site of replication and large nuclear inclusions are produced, the virion possesses hemagglutinin for at least one species of erythrocyte, and there is a marked resistance to heat.

Hallauer et al.,[21,22] over the 10-year period 1960—1970, tested 41 strains of six different human cell lines obtained from 19 laboratories, and recovered parvovirus from 36 of them. The majority of isolates belonged to one serotype which had a close relationship to porcine parvovirus but not to parvoviruses from other species. It was concluded that the agent had probably been transmitted to cell cultures by the trypsin used for cell dispersion. These workers commented that the continued use of contaminated cell cultures in various laboratories was not surprising since parvoviruses usually produce only insignificant cytopathic changes that may not be recognized as virus-induced. A cyclic pattern of viral activity was observed, in which subcultures of cells with production of virus were followed by apparently virus-free subcultures, although a trace of hemagglutinin might be detected in the latter. Storage of cells at low temperature tended to activate latent parvovirus infection when the cells were subsequently cultured.

A strain of parvovirus recovered in that study from the KB line of human cells was further investigated[23] in the same cell system. Intranuclear inclusions were noted, particularly when a multiplicity of infection of 1.5 to 2.0 TCID50 per cell was used, but the number of cells with characteristic inclusions never exceeded 6% of all the cells in

* All figures appear following page 58.

a monolayer culture. Inclusion formation reached a maximum at 24 hr after infection, but 48 hr later no inclusions could be detected. A high input multiplicity of virus could produce visible degeneration of cells, but low virus doses resulted only in some thinning of the cell sheet.

Croghan et al.[24] recovered a porcine parvovirus from a batch of commercially obtained trypsin by cultivating ultracentrifuged trypsin in freshly seeded cells of a swine testicle line. Presence of the virus was demonstrated by specific immunofluorescence, by the observation of nuclear inclusions in infected cells stained with May-Grunwald-Giemsa stain, and by hemagglutination tests. Uninoculated control cultures were negative for latent virus. It is possible that parvovirus could have survived the process of trypsin production in which a temperature of 65.5°C was maintained for 4 hr, because other studies on porcine parvovirus[25,26] have shown that the agent resists heating at 56°C for 30 min with no significant loss of titre, and that infectivity is not completely eliminated after holding at 70°C for 2 hr or at 56°C for 48 hr.

There is also a theoretical possibility that parvovirus may be introduced into cell cultures by way of bovine serum used in the medium. In 1961 Abinanti and Warfield[27] recovered a parvovirus from the GI tract of calves, using bovine embryonic kidney cell cultures for isolation of the agent. No cytopathic changes were observed initially, and the presence of virus was detected by hemadsorption with human and guinea pig erythrocytes. Storz and Bates,[28] in their study of 11 bovine parvovirus isolates, found that all known strains of the agent belonged to one serotype. Like other parvoviruses, the bovine agent has a predilection for actively growing cells and will only multiply in primary fetal bovine cells. Established cell lines of bovine, monkey, mouse, hamster, and human origin did not support replication of bovine parvovirus. It was shown that viremia followed oral inoculation of calves, and thus while the possibility remains that parvovirus may be found in bovine serum that is widely used in culture medium, the restricted host cell range of the virus makes it unlikely that it will present problems in cultures other than primary bovine fetal cells.

Feline panleukopenia (FPL) virus, which was first described in 1939 by Hammon and Enders,[29] is a parvovirus that occurs in all parts of the world in all members of the cat family.[30,31] It, too, requires rapidly dividing cells for optimal replication, and it produces nuclear inclusions in susceptible cells. As an illustration of nuclear inclusions produced by a parvovirus, we show in Figure 49 the effect of feline panleukopenia virus in cells of the CRFK line of feline renal cells.[32]

III. PAPOVAVIRUSES

The papovavirus group[33] includes the papilloma viruses of man and animals, the polyoma virus of mice, the vacuolating simian virus SV 40, and other DNA agents having similar morphology and general properties.

Papillomas (warts) in man are transmissible and contain particles with the morphology of papovaviruses. There is some evidence that there may be more than one type of human papilloma virus.[34,35] Histological examination of warts has revealed the presence of intranuclear inclusions and cytoplasmic masses in some warts.[36,37] Attempts at cultivation of human papilloma virus have, however, not met with unqualified success, although replication may occur in human fetal skin cultures and in mouse embryo cell cultures, accompanied by rounding and detachment of cells.[38] The papilloma viruses of animals have also presented difficulties in cultivation, but transformation of various cell types has been achieved with rabbit and bovine papilloma viruses.

Polyoma virus of mice causes degeneration of mouse embryo cell cultures, with small pyknotic cells increasing in number and detaching from the glass.[39] Plaques produced in monolayers of mouse embryo cells have been used to assay polyoma virus.[40]

The description of simian virus 40 in 1960 by Sweet and Hilleman[41] has led to the accumulation of a voluminous literature about this agent that occurs in nature in macaque monkeys, especially the rhesus monkey *Macaca mulatta.* The name vacuolating agent was given to it because cytoplasmic vacuolation in infected African green monkey kidney cells was the distinctive feature which led to its initial recognition. Such vacuolation is a convenient sign of infection, but it is less constant than the presence of basophilic nuclear inclusions which are always to be found in productive infections. Their detection is particularly important in macaque monkey or human cells where vacuolation may be very limited or absent. Figure 50 illustrates the fully developed fluffy basophilic masses in the nucleus of a baboon kidney cell. Occasionally a few well-defined vacuoles may be found in the nucleus in addition to the inclusions. Figure 51 shows, at a lower magnification, a monolayer of baboon kidney cells in which the majority of nuclei have basophilic inclusions, but only four cells show cytoplasmic vacuolation.

Hsiung and Gaylord[42] observed that in patas monkey kidney cultures SV 40 produced nuclear inclusions 2 to 3 days before cytoplasmic vacuolation appeared. In rhesus monkey kidney cells the nuclear changes were delayed and vacuolation did not occur. Miyamura and Kitahara[43] noted that at a high input multiplicity of SV 40 there appeared in African green monkey kidney cells an early cytoplasmic vacuolation as soon as 4 hr postinfection, before the synthesis of T or V antigens. This early vacuolation was distinctly coarse in appearance, differing from the finer vacuolation regarded as specific for SV 40 infection. At 10 to 20 hr after infection the early vacuolation reached a maximum and then declined until it had largely disappeared at 48 hr after infection when it began to be replaced by the finer and more characteristic vacuolation. SV 40 inactivated by ultraviolet irradiation was also capable of inducing the early vacuolation.

It was observed by Norkin[44] that in rhesus kidney cell cultures a multiplicity of 100 plaque-forming units of SV 40 resulted in a stable carrier state characterized by extensive viral proliferation with little, if any, cytopathic effect. Meyer et al.[45] noted that SV 40 produced visible cytopathic changes in primary African green monkey (*Cercopithecus*) kidney cells only after considerable quantities of virus had been produced. Furthermore, since small amounts of virus inoculum induced visible changes only in the 3rd to 5th weeks, cercopithecus kidney cultures used in diagnostic procedures had to be held for longer than 2 weeks. These workers found that the LLC-MK2 line of rhesus kidney cells was not sufficiently sensitive to detect consistently even $10^{2.6}$ TCID50/m*l* of SV 40, and vacuolation was not a prominent feature of infection in these cells.

SV 40 is capable of transforming various cells, including human diploid lung fibroblasts and human amnion cells. The term transformation was introduced by Koprowski et al.[46] to describe the morphological change undergone by cells that exhibited increased growth and karyotypic alteration following infection by SV 40.

In 1971 Padgett et al.[47] reported the isolation of a papovavirus from the brain of a case (J.C.) of progressive multifocal leucoencephalopathy (PML) in cultures of primary human fetal glial cells. Changes in the astrocytes about 30 days after inoculation with brain extract consisted of enlargement of cells with deeply staining bizarre nuclei that did not show inclusions when stained with hematoxylin and eosin. Further passage in mixed astrocyte and spongioblast cultures led to visible necrotic changes, and some cells contained compact nuclear inclusions which were eosinophilic or amphophilic. Electron microscopy showed the presence of particles resembling papovavirus in both nucleus and cytoplasm. The agent was inoculated into other types of human cells and into African green monkey kidney cells, but no cytopathic changes were produced.

A subsequent study[48] confirmed that the JC virus grew readily only in primary human fetal glial cell cultures, affecting both astrocytes and spongioblasts but chiefly the latter. Nuclear inclusions in spongioblasts were shown to contain DNA by Feulgen and acridine orange staining. The virus hemagglutinated human, guinea pig, and chicken erythrocytes, and the hemagglutination test was used to confirm the failure of JC virus to grow in a wide variety of cell cultures. Serological tests showed that JC virus had a minor relationship with SV 40. Takemoto et al.[49] successfully cultivated JC virus in primary human amnion cells for eight passages, but found that cellular changes comprising scattered swollen or bizarre-shaped cells were subtle and difficult to detect. The presence of JC viral antigen was confirmed by immunofluorescence, and intranuclear virions were observed by electron microscopy.

In 1972 Weiner et al.[50] reported the isolation of viruses closely related to SV 40 from the brains of two cases of PML. Cells cultured from brain biopsy material taken from one case were fused with primary African green monkey kidney (PAGMK) cells by means of inactivated Sendai virus. The heterokaryons showed no cytopathic changes, but homogenates of heterokaryons were inoculated into PAGMK cells, and after two successive passages cytoplasmic vacuolation and cell degeneration appeared. In subsequent passages these changes developed in 5 to 7 days. Virus was similarly recovered from a brain biopsy of the second patient, but the agent was also cultivated from homogenates of cultured brain biopsy cells that were inoculated into PAGMK cultures without the formation of heterokaryons. The two isolates were closely related to each other and to SV 40, as shown by neutralization tests. In order to exclude the possibility that a simian virus may have been introduced by the use of primary monkey kidney cells, a further study was carried out[51] in which a homogenate of stored frozen brain material from the first case was inoculated into cultures of primary human fetal brain cells. Subpassages at 28-day intervals were performed, and 18 days after the third passage there appeared cytopathic effects consisting of pyknosis and cell lysis. This isolate produced vacuolation and lysis of cells when subinoculated into African green monkey kidney cells, and it was confirmed serologically as identical to the original agent recovered from this patient.

The two SV 40-PML agents and the JC-PML virus differed from another papovavirus isolated by Gardner and her colleagues[52] from the urine of a patient (B.K.), a renal allograft recipient. Primary rhesus monkey kidney cultures inoculated with urine showed a gross cytopathic change after 18 days, but this could not be reproduced in serial passages during which virus failed to replicate. Vero cells inoculated with urine developed visible cytopathic changes 1 to 3 months later, showing darkening of the cytoplasm of cells with the appearance of scattered granular round cells on the surface of the monolayer. Complete destruction of the cell sheet followed after a further 4 weeks of slow development. Cultures of human embryo lung fibroblasts were inoculated with infective urine but did not support the growth of this agent. The BK virus was shown by electron microscopy to have the morphology of a papovavirus, and a low level of cross-reactivity with SV 40 was demonstrated by serological tests.

When cells in the urine of the patient BK were stained with hematoxylin and eosin, large single homogeneous basophilic intranuclear inclusions were seen, and by electron microscopy papovavirus particles were observed in cells in the urine.[53] Similar findings were obtained from urine samples taken from other renal allograft patients as well as from patients immunosuppressed for other reasons, and serological studies suggested that infection with BK virus in immunosuppressed patients may represent reactivation of a latent infection.

Lecatsas et al.[54] inoculated BK virus into primary cultures of human fetal fibroblasts and glial cells. Both cell types showed similar changes: after about 14 days the cell sheet began to appear roughened, and localized foci of swollen cells developed by the

24th day. Cytoplasmic vacuolation was present, particularly in glial cells. Stained coverslip cultures showed the presence of dense nuclear inclusions resembling those induced by other papovaviruses. Further information concerning the host cell range of BK virus is to be found in the review of the new human papovaviruses by Padgett and Walker.[55]

In 1963 we designated as SA 12 a virus recovered on one occasion only from primary vervet monkey kidney cells.[56] This agent produced nuclear changes similar to those of SV 40 but without cytoplasmic vacuolation, and it was not neutralized by SV 40 antiserum. In Figure 52 are illustrated the basophilic nuclear inclusions induced in vervet kidney cells by SA 12. Valis et al.[57] subsequently characterized it as a new simian papovavirus with the chacma baboon *Papio ursinus* as its probable natural host. SA 12 is not related serologically to other papovaviruses, but a capsid antigen common to agents of the SV 40-polyoma subgroup is present in cells infected with SA 12.

Other animal papovaviruses producing vacuolation of certain cell types include the agent isolated from the kidneys of stump-tailed macaque monkeys (*Macaca speciosa*),[58,59] as well as the virus recovered by Hartley and Rowe from cottontail rabbits.[60] It is probable that in due course more papovaviruses will be isolated from human and animal sources, and the use of stained coverslip cultures should greatly facilitate early recognition of such agents.

IV. ADENOVIRUS

Adenoviruses[61,62] were first detected by Rowe et al.[63] when human adenoid explant cultures showed rounding of cells followed by complete degeneration. Subinoculation of fluid harvested from these cultures into human embryonic cells resulted in similar degeneration. Infected human embryo tracheal epithelium cultures stained with hematoxylin and eosin showed the presence of nuclear inclusions that were usually basophilic but occasionally eosinophilic.

The cytopathic changes induced by adenoviruses are varied and spectacular, and a single infected cell culture may contain a diversity of changes rarely encountered with other viruses. The kind of fixative used for stained preparations influences the morphological appearance of the nuclear changes.[64] Much has been written about the successive changes observed in infected cells, and differences between the effects produced by various adenovirus serotypes have been reviewed by Ginsberg and Dingle.[65] The correlation between morphological and biochemical events in productive cells has been outlined by Ginsberg:[66] early eosinophilic inclusions represent stored early viral proteins, while the central basophilic mass formed later comprises excess viral DNA and capsid proteins.

It was observed by Pereira[67] and by Rowe and his colleagues[68] that there existed a factor responsible for the early rounding of adenovirus-infected cells and for their detachment from the glass, which was separable from the virus particles, nondialysable, and sensitive to trypsin. These early changes occurred within a few hours following infection and could be reversed by changing the medium. This toxic effect is due to the penton component of the adenovirus virion acting on the surface of the cell. Subsequent nuclear changes are associated with partial or complete replication of virus inside the nucleus.

Studies by Morgan and his colleagues[69-71] revealed the presence of protein paracrystals in adenovirus-infected cells. These structures, which occurred in the nucleus but did not contain virus particles, were also found in the cytoplasm after the nucleus had become disrupted. The paracrystals are readily produced by adenovirus type 5, but other serotypes can do so as well.[72,73] Figure 63 illustrates paracrystals in the nucleus of a human amnion cell infected with adenovirus type 5. It has been postulated that these bodies represent accumulations of core or capsid viral proteins.[74-76] Although the paracrystals appear hexagonal in some cross sections, they may in fact be rectangular.[77]

The earliest nuclear changes in adenovirus-infected cells take the form of small eosinophilic inclusions (Figure 53). At this stage infectious virus has not yet been replicated.[78] Other early and highly distinctive changes comprise multiple discrete and well-defined inclusions, each of which consists of a basophilic center surrounded by a pale area enclosed within a more deeply-staining border (Figure 54). Subsequently a moruloid basophilic inclusion fills the nucleus and contains multiple vacuoles that may distort the nuclear membrane (Figure 55). The morula then contracts into a dense central mass around which are relatively few vacuoles apparently formed by the fusion of smaller vacuoles. The material inside the vacuoles is eosinophilic or amphophilic. Further fusion of vacuoles results in a ring-like zone around the central morula, and in this zone is contained the coalesced contents of the vacuoles. In Figure 56 we illustrate in one cell the penultimate stage with large vacuoles, and in another cell the final stage with a ring of amphophilic material lining the nuclear membrane. Cytoplasmic inclusions are present in the latter cell: at this late stage well-defined and deeply staining cytoplasmic bodies are induced by some adenoviruses, although viral replication takes place only in the nucleus. Figure 57 also illustrates the late stage with cytoplasmic inclusions. By contrast, we present in Figure 63 the larger and paler eosinophilic cytoplasmic inclusions produced by adenovirus type 5 in human amnion cells, which are not present in simian cells infected with this agent.

The range of adenovirus-induced changes is extensive, varying with the strain of virus and the type of cell. Figures 58 through 62 illustrate some of the less common nuclear changes, all of which are variations on the same theme but indicate strain differences that may warrant special attention. Figure 60 shows basophilic bodies in the zone adjacent to the nuclear membrane, while Figures 61 and 62 illustrate eosinophilic structures in the same area. Adenovirus infection of cell cultures occasionally leads to the formation of multinucleated giant cells which are not syncytia but result from cytophagocytosis. Figure 64 shows an infected cell being ingested during the process of giant cell formation.

Human adenoviruses replicate in a variety of primary human cells, particularly embryo kidney, as well as in human cell lines such as KB, HEp-2, and HeLa, although some sublines of HeLa cells have been found to differ in their sensitivity.[79] On the other hand, human adenoviruses do not multiply significantly in simian cells in spite of the fact that nuclear morphological changes characteristic of adenovirus infection are found in such infected cells. The inclusions in abortively infected monkey cells are Feulgen negative and do not contain viral capsid protein,[80,81] but they do contain adenovirus tumor antigen.[82]

Rabson et al.[83] found that adenovirus type 12 inoculated into African green monkey kidney cells produced cell rounding and nuclear changes but did not replicate, whereas simultaneous infection with SV 40 promoted growth of the adenovirus. Goulet et al.[84] inoculated a number of human adenovirus serotypes into both primary human embryo kidney cells and primary African green monkey kidney cells, and showed that adaptation of five serotypes to growth in simian cells could occur provided that the initial isolation and 1 to 5 subsequent passages was performed in human cells before subinoculation at high dosage in monkey cells were carried out. Prolonged incubation of the simian cell cultures at 37°C was necessary, together with careful observation for the development of foci of adenovirus infection. The adapted variant strains were antigenically identical to their parent strains. Selection of a variant of human adenovirus type 2 capable of unrestricted replication in simian cells was achieved by Klessig[85] through treatment of high-titered adenovirus stock with nitrous acid. This variant was further characterized[86] by means of restriction endonuclease and hybridization analyses to exclude the possibility of contamination of its genome with SV 40.

O'Conor et al.[87] showed by electron microscopy that simultaneous infection of simian cells with SV 40 and human adenovirus type 12 led to double infection in the nuclei of about 1% of the cells, suggesting that phenotypic mixing of the viruses might be possible. Subsequent studies[88-91] indicated that genetic hybrids of SV 40 and adenovirus are produced in simian cells, and that the combined genetic material is enclosed within the capsid of the helper adenovirus. Replication of the hybrid known as PARA (Particle Aiding, and aided by, Replication of Adenovirus)[92] is accompanied in the same cell by replication of intact adenovirus, both agents being mutually dependent for replication during lytic infection of simian cells.

Double infection with the adenovirus-associated parvovirus (AAV) and adenovirus is necessary for the regular replication of AAV, but the presence of AAV may cause abortive infection with adenovirus and interfere with plaque production.[93,94]

The oncogenic potential of a number of human and simian adenoviruses is well-known. In Figure 65 we illustrate a focus of neonatal hamster cells transformed in vitro by the simian adenovirus SA 7,[95,56] which is highly oncogenic for newborn hamsters[96] and which differs from other simian adenoviruses in its lack of hemagglutinin.[97]

A number of investigators have observed that occasionally stools taken from children with diarrhea contain adenovirus-like particles that cannot be cultivated in conventional cell culture systems. Gary et al.[98] detected such noncultivable particles in the stools of 13 children by means of electron microscopy. Using counterimmunoelectrophoresis and immunofluorescence with antihexon serum they showed that the particles were adenovirus, capable of infecting cells but unable to undergo effective replication in the human and simian cells used. It is clear that the virus can multiply at some site in the body, but a suitable cell system for in vitro cultivation remains to be found.

V. HERPESVIRUSES

Herpesviruses are harbored by a number of animal species, and the cytopathic changes produced by them have certain features in common. In some instances, however, distinctive characteristics can be recognized in infected cell cultures, and we shall illustrate some of these differences. As the name implies, herpesvirus infection creeps from cell to adjacent cell, and in monolayer cultures enlargement of established foci rather than the formation of new foci is commonly seen.

A. Herpes Simplex Virus

Herpes simplex virus, or *Herpesvirus hominis*, can be cultivated in a wide range of cell types of which human embryo and rabbit kidney cells have been preferred by many workers. Nuclear inclusions are produced by all strains of the virus, but the formation of multinucleated cells is variable, depending on the genetic constitution of the strain, the cell type used for cultivation, and the temperature of incubation.[99-101] Some strains chiefly induce rounding of individual cells, while others characteristically produce syncytia. The presence of only occasional syncytia can be due to the coexistence of two strains separable by the plaque technique[102] which can also be used to separate large- and small-syncytium-producing variants.[103] The process by which multinucleated cells are induced by herpes simplex virus is usually one of syncytium formation rather than of cytophagocytosis. The factor mediating fusion has been identified as a glycoprotein separable from the infective particle by sucrose gradient centrifugation[104] or by ultraviolet irradiation of the virus.[105] This factor appears to be under the control of another glycoprotein which acts as a suppressor of fusion.[106] The antimetabolite 2-deoxy-D-glucose inhibits cell fusion by interfering with the glycosylation of glycoprotein.[107]

Normally herpes-induced cell fusion is an endogenous event proceeding from within the cell at a stage when little or no infectious virus is being produced.[108]

The cytochemical study by Love and Wildy[109] showed that in HeLa cells infected with herpes simplex virus early changes comprised the formation within the nucleoli of small ribonucleoprotein bodies which were then extruded into the nucleoplasm where they enlarged and aggregated prior to the appearance of the large DNA-containing inclusion body. Crouse et al.[110] observed that in the chorioallantoic cells of infected chick embryos the earliest identifiable nuclear inclusion was a Feulgen-positive body which almost completely filled the nucleus and stained bluish with hematoxylin and eosin. The chromatin was marginated and remained Feulgen-positive, whereas the inclusion became Feulgen-negative as it aged. Kristensson et al.[111] studied the ultrastructural and histochemical properties of the large inclusions in neurones of mice infected with herpes simplex virus, and noted that the inclusions contained relatively few virus particles, some DNA, and strongly basic proteins that were responsible for the characteristic eosinophilia of the inclusions. Morgan and his colleagues[112,113] have pointed out that it is virtually impossible to visualize with the electron microscope the inclusion body seen by light microscopy. Furthermore, since it would take approximately 300 sections averaging 40 nm in thickness to examine the entire contents of a normal cell, events in the swollen and irregularly shaped infected nucleus would be still more difficult to assess, and structures encountered at one level by electron microscopy cannot be assumed to reflect the state of the entire nucleus.

In Figure 66 are shown herpes simplex-infected monkey kidney cells illustrating the early basophilic inclusion filling the nucleus with its marginated chromatin. The presence of small deeply basophilic bodies scattered throughout the early inclusion is characteristic of herpes simplex and allied herpes infections. Two cells show the late contracted eosinophilic inclusion.

In Figure 67 we illustrate a focus of herpes simplex-infected rhesus kidney cells where most of the cells are rounded and remain single, only one small multinucleated cell being present. Figure 68 shows a focus of infection in a human fetal lung cell culture stained with fluorescein.

The oncogenic transformation of hamster embryo fibroblasts by ultraviolet-irradiated *Herpesvirus hominis* type 2 has been documented by Duff and Rapp.[114] A cell line derived from an in vitro transformed focus contained two cell types: a predominant undifferentiated cell, and an enlarged cell. The cells contained herpes specific antigens but no complete virus particles.

B. Other Herpesviruses

Other herpesviruses producing cytopathic changes comprising syncytia and nuclear inclusions somewhat resembling those of herpes simplex virus include *Herpesvirus simiae* or the B virus of monkeys. This agent was first isolated by Sabin and Wright[115] from the spleen, brain, and cord of a patient (W.B.) who had died following a rhesus monkey bite. The virus occurs naturally in several species of Asian macaque monkeys, and evidence of infection has been found in over 70% of adult wild rhesus monkeys in some parts of India.[116] Macaque monkeys have been widely used in research, and by 1973 24 human cases of B virus encephalitis had been recorded, of which 18 were fatal.[117] The virus can be cultivated in human, simian, rabbit, and chick embryo cells with the production of herpetic nuclear inclusions and syncytia. Reissig and Melnick[118] correlated the ultramicroscopic appearances of B virus-infected cells with changes seen by light microscopy and found, as mentioned earlier, that osmium fixation followed by hematoxylin and eosin staining failed to demonstrate clearly nuclear inclusions.

Variants of B virus have been selected by plaque size in chick embryo cell cultures.[119] Two variants were distinguished in a strain studied by Falke:[120] one produced large syncytia, and the other caused rounding of cells without syncytium formation. This distinction was retained when they were cultivated in several human and simian cells, except that the variant producing single cell rounding induced syncytia in HeLa cells but reverted to rounding when passaged again in KB cells. The two variants had the same pathogenicity for animals. Hull has noted[121] that in simian kidney cells B virus produces large multinucleated cells, whereas in rabbit kidney cultures it causes single cell rounding without enlargement.

A herpesvirus closely related to B virus, designated SA 8, was isolated by us from simians in Africa.[56,122,123] This agent causes lesions identical to those of B virus when inoculated into the central nervous system of monkeys, and it paralyzes rabbits after intradermal inoculation although to a lesser extent than B virus. In simian kidney cell cultures SA 8 induces widespread syncytia, one of which is illustrated in Figure 69 where the early basophilic inclusions predominate. The distribution of chromatin granules throughout the early inclusions is characteristic of infection with this group of herpesviruses. In the syncytium a few eosinophilic contracted late inclusions are present. Several nuclei have vacuoles reminiscent of adenovirus infection, but these never progress through the sequence of events outlined in Chapter 6,IV for adenovirus. SA 8 also produces small and rather indistinct eosinophilic cytoplasmic bodies.

Another African simian herpesvirus, SA 15, was isolated by us on one occasion from an uninoculated vervet monkey kidney cell culture[56] and several times from throat swabs taken from baboons.[124] Figure 70 shows the distinctive type of syncytium produced by this agent in vervet kidney cells. There is marked variation in size of the infected nuclei, and nuclear vacuoles containing eosinophilic material are present. These vacuoles illustrate the resemblance between herpesvirus and adenovirus cytopathic changes that may occasionally be seen, but SA 15 has been shown by electron microscopy to be a herpesvirus.

In contrast we show in Figure 71 a syncytium produced in Vero cells by a herpesvirus isolated from a South American spider monkey. Here the cellular changes conform more closely to those induced by herpes simplex virus, with well-contracted eosinophilic inclusions.

A rabbit herpesvirus may be encountered as a latent infection in primary rabbit kidney cultures.[125,126] The presence of this agent can be readily recognized by focal syncytium formation and the typical herpetic nuclear inclusions.

C. Cytomegalovirus

In several respects the cytopathic changes produced by cytomegaloviruses may be distinguished from those induced by the herpesviruses discussed above. Some of these features, as seen in vervet monkey kidney cells, are illustrated in Figures 74 and 75. Early nuclear inclusions usually take the form of one or two eosinophilic bodies which enlarge or coalesce to form a single inclusion that is frequently elongated. The prominent nucleoli marginate with the chromatin along the nuclear membrane. The texture of the inclusion material remains fairly constant, and the large basophilic inclusion with chromatin granules in it that is seen in herpes simplex-infected cells is not present. Inclusion material tends to be amphophilic and stains a deep purple-red with hematoxylin and eosin, while the cytoplasm stains a deep red. Small basophilic cytoplasmic bodies may be seen in some human fibroblasts infected with cytomegalovirus. The nucleus may be displaced laterally by a pale cytoplasmic zone. Waner and Weller[127,128] inoculated human strains of cytomegalovirus into the Vero line of monkey kidney cells and noted the development of a paranuclear granular body in the absence of nuclear inclusions.

Figures 74 and 75 also show the formation of multinucleated cells as the result of virally stimulated cytophagocytosis. The outline of the newly ingested cell remains distinct until the process of engulfment has been completed. Some writers refer to such giant cells as syncytia, but we prefer to preserve the distinction between cytophagocytosis and syncytium formation through fusion of plasma membranes as a useful guide in assessing cytopathic changes induced by different viruses. Syncytium formation in this sense may contribute in some degree to the production of cytomegalic giant cells, but cytophagocytosis is the salient characteristic.

Diosi et al.[129] cultured cytomegalovirus strains from ground squirrels, dogs, and humans in primary embryonic skin-muscle or adult kidney cells derived from the host species, and observed that multinucleated cells formed by cytophagocytosis resulted from infection with all strains except five isolated from human throats. These five oral strains were further compared[130] in primary human embryonic skin-muscle cells with three strains of the virus obtained from the human female genital tract. The oral strains produced necrosis spreading rapidly throughout the entire cell sheet within 7 to 9 days without multinucleated cell formation. The genital strains, on the other hand, grew more slowly and produced giant cells by cytophagocytosis. This difference between the oral and the genital strains persisted during three cell culture passages and was taken as a reflection of inherent characteristics of the strains.

Cytomegaloviruses display a preference for cells of the host species, but this is not absolute. Simian virus can be cultivated in human embryo fibroblasts,[131] and the human virus has been grown in the Vero line of monkey kidney cells.[127,128] In the latter instance, virus was not recovered from the fluid harvested from infected cultures, but co-cultivation of the cells with human fibroblasts revealed the presence of replicating virus.

For some time it was believed that human cytomegalovirus could only be cultivated in nonepithelial cells, although the presence of inclusions in epithelial cells in various tissues in the body had been noted in histological sections. Figures 72 and 73 illustrate the distinctive appearance of an infected cell in the salivary gland of a vervet monkey, the large eosinophilic inclusion lying within the ring of marginated chromatin, and the deeply staining cytoplasm being easily detectable even at the lower magnification. The simian virus occurs frequently as a latent agent in the epithelial cells of monkey kidney cultures, and there is now ample evidence that the human virus can be cultivated in epithelial cells such as those derived from the thyroid[132] or amnion.[133,134] Different epithelial cell types may, however, vary in their sensitivity; for instance, ectocervical cells of the uterine epithelium are refractory, whereas endocervical cells are sensitive to human cytomegalovirus.[135] St. Jeor and Rapp[136] converted nonpermissive human embryonic kidney cells to a permissive state in which infectious virus and cytopathic changes were produced by treating the cells with 5-iodo-2-deoxyuridine prior to infection. Michelson-Fiske et al.[137] carried out electron microscopic studies on cultures of human embryonic lung cells trypsinized to remove fibroblasts and leave epithelial cells which were then infected with a strain of human cytomegalovirus. Viral replication occurred in the absence of ultrastructural lesions, and it was suggested that epithelial cells might have a role in the steady state of cytomegalovirus infection in vivo when there is long-term excretion of virus in urine or saliva.

Early workers mistakenly regarded cytomegalovirus-infected cells seen in histological sections as protozoa. In 1910 Smith and Weidman proposed the name *Endamoeba mortinatalium* for the supposed organisms seen in the viscera of a stillborn fetus, and they suggested that aberrant mouth amoebae might account for the presence of the organisms in the salivary glands of other cases.[138] Goodpasture and Talbot[139] recognized the nonprotozoal nature of the enlarged cells and described the development of eosinophilic intranuclear bodies which, they recalled, had also been observed previously in cutaneous lesions of varicella. Not having any evidence to support the concept of an infection, they suggested that the condition be called cytomegalia. After strains of the virus had been isolated from man and animals, Weller et al.[140] proposed the name cytomegalovirus for this group of agents.

D. Varicella-Zoster Virus

Varicella-zoster virus was first serially cultivated by Weller in human embryonic skin-muscle or foreskin cell culturees.[141] It has since been shown that the typical cytopathic changes can be produced in other human and animal cells, some of which support only abortive infection and serve chiefly as indicator systems for the presence of virus.[142-144] Plaque assays have been carried out in several cell systems including human embryonic lung fibroblasts,[145] human melanoma cells,[146] and mink lung cells.[147] Various cell types susceptible to varicella-zoster virus have been listed by Taylor-Robinson and Caunt.[148]

Vesicle fluid contains infectious cell-free virus, but most of the cell cultures used to propagate the agent liberate little or no free virus into the surrounding medium. Transfer of viable infected cells has usually been employed in serial passages, and for this purpose trypsinized infected cells can be used.[149] Caunt and Taylor-Robinson observed that ultrasonication of vesicle fluid resulted in reduction of infectivity, but similar treatment of infected human thyroid cells released significant amounts of undamaged virus.[150] Roller tube cultures may increase the number of foci by distributing infected cells. Other cultural conditions such as composition of the medium can influence viral growth.[144] Several workers have noted that at incubation temperatures below 37°C virus production is slower although not necessarily reduced in final titer;[148] but Grose and Brunell[146] found that for cultivation in human melanoma cells a temperature of 32°C was preferable.

The cytopathic effects of varicella-zoster virus include the production of nuclear inclusions and the formation of syncytia. At an early stage, basophilic material filling the nucleus and having chromatin granules in it as seen in herpes simplex infection may occur in some cultures. The later eosinophilic inclusions are usually multiple and characteristically have an irregular shape, as shown in Figures 77 and 78. The eosinophilic material has a finely granular texture and does not, as a rule, stain as deeply as that produced by cytomegalovirus. Nucleoli are prominent and tend to lie between the inclusions. The marginated chromatin may be finely granular rather than clumped. Multinucleated cells of the syncytial type are formed by some strains of the virus; and their production may also be temperature-dependent, with larger syncytia being formed at lower temperature.[146]

Several outbreaks of vesicular disease in monkeys have been caused by strains of a virus variously called Liverpool vervet virus,[151] patas herpes virus,[152] Delta herpes virus,[153] or Medical Lake macaque herpes virus.[154] This agent produces nuclear inclusions and multinucleated cells in certain human and simian cell cultures of which the Vero line of monkey kidney cells has most often been used. The strains of this virus are very closely related to each other and also have a serological relationship with varicella-zoster virus but are not identical to it.[155,156]

REFERENCES

1. **Cowdry, E. V. and Kitchen, S. F.,** Intranuclear inclusions in Yellow Fever, *Am. J. Hyg.*, 11, 227, 1930.
2. **Smith, H. H. and Theiler, M.,** The adaptation of unmodified strains of Yellow Fever virus to cultivation in vitro, *J. Exp. Med.*, 65, 801, 1937.

3. **Doherty, R. L.,** Effects of Yellow Fever (17D) and West Nile viruses on the reactions of human appendix and conjunctiva cells to several other viruses, *Virology*, 6, 575, 1958.
4. **Hallauer, C.,** Zuchtung von Gelbfiebervirus in menschlichen explantaten, *Arch. Gesamte Virusforsch.*, 9, 428, 1960.
5. **Buckley, S. M.,** Applicability of the HeLa (Gey) strain of human malignant epithelial cells to the propagation of arboviruses, *Proc. Soc. Exp. Biol. Med.*, 116, 354, 1964.
6. **Buckley, S. M.,** Propagation, cytopathogenicity, and hemagglutination-hemadsorption of some arthropod-borne viruses in tissue culture, *Ann. N.Y. Acad. Sci.*, 81, 172, 1959.
7. **Karabatsos, N. and Buckley, S. M.,** Susceptibility of the baby-hamster kidney-cell line (BHK-21) to infection with arboviruses, *Am. J. Trop. Med. Hyg.*, 16, 99, 1967.
8. **Mussgay, M., Enzmann, P. J., Horzinek, M. C., and Weiland, E.,** Growth cycle of arboviruses in vertebrate and arthropod cells, *Prog. Med. Virol.*, 19, 257, 1975.
9. **Borisova, S. M., Demidova, S. A., Monastyreva, L. A., Levina, D. S., Zakaryan, V. A., Veselovskaya, O. V., Deryabin, P. G., Ratushkina, L. S., Gavrilov, V. I., and Lvov, D. K.,** Use of duck embryo cell cultures for the study of arbovirus reproduction, *Acta Virol. Engl. Ed.*, 19, 432, 1975.
10. **Yershov, F. I. and Zhdanov, V. M.,** Relationship between group A arbovirus reproduction and cytochemical changes in infected cells, *Acta Virol. Engl. Ed.*, 12, 136, 1968.
11. **Rajcani, J. and Gresikova, M.,** Multiplication of Tribec arbovirus in L cells, *Acta Virol. Engl. Ed.*, 14, 175, 1970.
12. **Karmysheva, V. Y., Semashko, I. V., Borisov, V. M., and Chumakov, M. P.,** Comparative study of cytopathology produced by some Kemerovo group viruses, *Acta Virol. Engl. Ed.*, 17, 479, 1973.
13. **David-West, T. S.,** Propagation and plaquing of Dugbe virus (an ungrouped Nigerian arbovirus) in various mammalian and arthropod cell lines, *Arch. Gesamte Virusforsch.*, 44, 330, 1974.
14. **Taylor, R. M., Hurlbut, H. S., Work, T. H., Kingston, J. R., and Frothingham, T. E.,** Sindbis virus: a newly recognized arthropod-transmitted virus, *Am. J. Trop. Med. Hyg.*, 4, 844, 1955.
15. **Malherbe, H., Strickland-Cholmley, M., and Jackson, A. L.,** Sindbis virus infection in man. Report of a case with recovery of virus from skin lesions, *S. Afr. Med. J.*, 37, 547, 1963.
16. **Ben-Ishai, Z., Goldblum, N., and Becker, Y.,** The intracellular site and sequence of Sindbis virus replication, *J. Gen. Virol.*, 2, 365, 1968.
17. **Veckenstedt, A. and Wagner, M.,** Virological and immunofluorescent studies on the multiplication of Sindbis virus in L cells, *Arch. Gesamte Virusforsch.*, 42, 144, 1973.
18. **Coackley, W.,** The effect of recently isolated strains of Rift Valley Fever virus on lamb testis cell cultures, *J. Pathol. Bacteriol.*, 86, 530, 1963.
19. **Bachmann, P. A., Hoggan, M. D., Kurstak, E., Melnick, J. L., Pereira, H. G., Tattersall, P., and Vago, C.,** Parvoviridae: second report, *Intervirology*, 11, 248, 1979.
20. **Siegl, G.,** The Parvoviruses, *Virol. Monogr.*, 15, 4, 1976.
21. **Hallauer, C., Kronauer, G., and Siegl, G.,** Parvoviruses as contaminants of permanent human cell lines. I. Virus isolation from 1960—1970, *Arch. Gesamte Virusforsch.*, 35, 80, 1971.
22. **Hallauer, C., Siegl, G., and Kronauer, G.,** Parvoviruses as contaminants of permanent human cell lines. III. Biological properties of the isolated viruses, *Arch. Gesamte Virusforsch.*, 38, 366, 1972.
23. **Siegl, G., Hallauer, C., and Novak, A.,** Parvoviruses as contaminants of permanent human cell lines. IV. Multiplication of KBSH virus in KB cells, *Arch. Gesamte Virusforsch.*, 36, 351, 1972.
24. **Croghan, D. L., Matchett, A., and Koski, T. A.,** Isolation of porcine parvovirus from commercial trypsin, *Appl. Microbiol.*, 26, 431, 1973.
25. **Mayr, A., Bachmann, P. A., Siegl, G., Mahnel, H., and Sheffy, B. E.,** Characterization of a small porcine DNA virus, *Arch. Gesamte Virusforsch.*, 25, 38, 1968.
26. **Cartwright, S. F., Lucas, M., and Huck, R. A.,** A small haemagglutinating porcine DNA virus. I. Isolation and properties, *J. Comp. Pathol.*, 79, 371, 1969.
27. **Abinanti, F. R. and Warfield, M. S.,** Recovery of a hemadsorbing virus (HADEN) from the gastrointestinal tract of calves, *Virology*, 14, 288, 1961.
28. **Storz, J. and Bates, R. C.,** Parvovirus infections in calves, *J. Am. Vet. Med. Assoc.*, 163, 884, 1973.
29. **Hammon, W. D. and Enders, J. F.,** A virus disease of cats, principally characterized by aleucocytosis, enteric lesions and the presence of intranuclear inclusion bodies, *J. Exp. Med.*, 69, 327, 1939.
30. **Scott, F. W., Csiza, C. K., and Gillespie, J. H.,** Feline viruses. IV. Isolation and characterization of Feline Panleukopenia virus in tissue culture and comparison of cytopathogenicity with feline picornavirus, herpesvirus, and reovirus, *Cornell Vet.*, 60, 165, 1970.
31. **Gillespie, J. H. and Scott, F. W.,** Feline viral infections, *Adv. Vet. Sci. Comp. Med.*, 17, 163, 1973.
32. **Crandell, R. A., Fabricant, C. G., and Nelson-Rees, W. A.,** Development, characterization, and viral susceptibility of a feline (Felis catus) renal cell line (CRFK), *In Vitro*, 9, 176, 1973.
33. **Melnick, J. L.,** Papova virus group, *Science*, 135, 1128, 1962.
34. **Orth, G., Favre, M., and Croissant, O.,** Characterization of a new type of human papillomavirus that causes skin warts, *J. Virol.*, 24, 108, 1977.

35. **Gissmann, L., Pfister, H., and Zur Hausen, H.,** Human papilloma viruses (HPV): characterization of four different isolates, *Virology,* 76, 569, 1977.
36. **Strauss, M. J., Bunting, H., and Melnick, J. L.,** Virus-like particles and inclusion bodies in skin papillomas, *J. Invest. Dermatol.,* 15, 433, 1950.
37. **Strauss, M. J., Bunting, H., and Melnick, J. L.,** Eosinophilic inclusion bodies and cytoplasmic masses in verrucae, *J. Invest. Dermatol.,* 17, 209, 1951.
38. **Oroszlan, S. and Rich, M. A.,** Human wart virus: in vitro cultivation, *Science,* 146, 531, 1964.
39. **Eddy, B. E., Stewart, S. E., and Berkeley, W.,** Cytopathogenicity in tissue cultures by a tumor virus from mice, *Proc. Soc. Exp. Biol. Med.,* 98, 848, 1958.
40. **Sachs, L., Fogel, M., and Winocour, E.,** In vitro analysis of a mammalian tumour virus, *Nature (London),* 183, 663, 1959.
41. **Sweet, B. H. and Hilleman, M. R.,** The vacuolating virus S.V.40, *Proc. Soc. Exp. Biol. Med.,* 105, 420, 1960.
42. **Hsiung, G. D. and Gaylord, W. H.,** The vacuolating virus of monkeys. I. Isolation, growth characteristics, and inclusion body formation, *J. Exp. Med.,* 114, 975, 1961.
43. **Miyamura, T. and Kitahara, T.,** Early cytoplasmic vacuolization of African green monkey kidney cells by SV 40, *Arch. Virol.,* 48, 147, 1975.
44. **Norkin, L. C.,** Effect of input multiplicity on the establishment of simian virus 40 persistent infections in rhesus monkey kidney cells, *Infect. Immun.,* 18, 868, 1977.
45. **Meyer, H. M., Hopps, H. E., Rogers, N. G., Brooks, R. E., Bernheim, B. C., Jones, W. P., Nisalak, A., and Douglas, R. D.,** Studies on simian virus 40, *J. Immunol.,* 88, 796, 1962.
46. **Koprowski, H., Ponten, J. A., Jensen, F., Ravdin, R. G., Moorhead, P., and Saksela, E.,** Transformation of cultures of human tissue infected with simian virus SV 40, *J. Cell. Comp. Physiol.,* 59, 281, 1962.
47. **Padgett, B. L., Walker, D. L., ZuRhein, G. M., Eckroade, R. J., and Dessel, B. H.,** Cultivation of papova-like virus from human brain with progressive multifocal leucoencephalopathy, *Lancet,* 1, 1257, 1971.
48. **Padgett, B. L., Rogers, C. M., and Walker, D. L.,** JC virus, a human polyomavirus associated with progressive multifocal leukoencephalopathy: additional biological characteristics and antigenic relationships, *Infect. Immun.,* 15, 656, 1977.
49. **Takemoto, K. K., Howley, P. M., and Miyamura, T.,** JC human papovavirus replication in human amnion cells, *J. Virol.,* 30, 384, 1979.
50. **Weiner, L. P., Herndon, R. M., Narayan, O., Johnson, R. T., Shah, K., Rubinstein, L. J., Preziosi, T. J., and Conley, F. K.,** Isolation of virus related to SV 40 from patients with progressive multifocal leukoencephalopathy, *N. Engl. J. Med.,* 286, 385, 1972.
51. **Weiner, L. P., Herndon, R. M., Narayan, O., and Johnson, R. T.,** Further studies of a simian virus 40-like virus isolated from human brain, *J. Virol.,* 10, 147, 1972.
52. **Gardner, S. D., Field, A. M., Coleman, D. V., and Hulme, B.,** New human papovavirus (BK) isolated from urine after renal transplantation, *Lancet,* 1, 1253, 1971.
53. **Coleman, D. V.,** The cytodiagnosis of human polyomavirus infection, *Acta Cytol.,* 19, 93, 1975.
54. **Lecatsas, G., Prozesky, O. W., and Scheepers, F.,** The cytopathology and development of a human polyoma virus (B.K.), *Arch. Gesamte Virusforsch.,* 45, 319, 1974.
55. **Padgett, B. L. and Walker, D. L.,** New human papovaviruses, *Prog. Med. Virol.,* 22, 1, 1976.
56. **Malherbe, H., Harwin, R., and Ulrich, M.,** The cytopathic effects of vervet monkey viruses, *S. Afr. Med. J.,* 37, 407, 1963.
57. **Valis, J. D., Newell, N., Reissig, M., Malherbe, H., Kaschula, V. R., and Shah, K. V.,** Characterization of SA 12 as a simian virus 40-related papovavirus of chacma baboons, *Infect. Immun.,* 18, 247, 1977.
58. **Rangan, S. R., Lowrie, R. C., Roberts, J. A., Johnston, P. B., and Warrick, R. P.,** Virus from Stumptailed monkey (Macaca arctoides) kidney cultures, *Lab. Anim. Sci.,* 24, 211, 1974.
59. **Reissig, M., Kelly, T. J., Daniel, R. W., Rangan, S. R., and Shah, K. V.,** Identification of the Stumptailed macaque virus as a new papovavirus, *Infect. Immun.,* 14, 225, 1976.
60. **Hartley, J. W. and Rowe, W. P.,** New papovavirus contaminating Shope papillomata, *Science,* 143, 258, 1964.
61. **Enders, J. F., Bell, J. A., Dingle, J. H., Francis, T., Hilleman, M. R., Huebner, R. J., and Payne, A. M.,** "Adenoviruses": group name proposed for new respiratory-tract viruses, *Science,* 124, 119, 1956.
62. **Norrby, E., Bartha, A., Boulanger, P., Dreizin, R. S., Ginsberg, H. S., Kalter, S. S., Kawamura, H., Rowe, W. P., Russell, W. C., Schlesinger, R. W., and Wigand, R.,** Adenoviridae, *Intervirology,* 7, 117, 1976.
63. **Rowe, W. P., Huebner, R. J., Gilmore, L. K., Parrott, R. H., and Ward, T. G.,** Isolation of a cytopathogenic agent from human adenoids undergoing spontaneous degeneration in tissue culture, *Proc. Soc. Exp. Biol. Med.,* 84, 570, 1953.

64. **Armstrong, J. A. and Pereira, H. G.**, Significance of cytopathic effects observed during the growth of adenovirus, *Exp. Cell Res.*, 21, 144, 1960.
65. **Ginsberg, H. S. and Dingle, J. H.**, The adenovirus group, in *Viral and Rickettsial Infections of Man*, 4th ed., Horsfall, F. L. and Tamm, I., Eds., Lippincott, Philadelphia, 1965, chap. 40.
66. **Ginsberg, H. S.**, The biochemical basis of adenovirus cytopathology, in *Infectious Agents and Host Reactions*, Mudd, S., Ed., W. B. Saunders, Philadelphia, 1970, chap. 20.
67. **Pereira, H. G.**, A protein factor responsible for the early cytopathic effect of adenoviruses, *Virology*, 6, 601, 1958.
68. **Rowe, W. P., Hartley, J. W., Roizman, B., and Levy, H. B.**, Characterization of a factor formed in the course of adenovirus infection of tissue cultures causing detachment of cells from glass, *J. Exp. Med.*, 108, 713, 1958.
69. **Morgan, C., Godman, G. C., Rose, H. M., Howe, C., and Huang, J. S.**, Electron microscopic and histochemical studies of an unusual crystalline protein occurring in cells infected by type 5 adenovirus. Preliminary observations, *J. Biophys. Biochem. Cytol.*, 3, 505, 1957.
70. **Morgan, C., Godman, G. C., Breitenfeld, P. M., and Rose, H. M.**, A correlative study by electron and light microscopy of the development of type 5 adenovirus. I. Electron microscopy, *J. Exp. Med.*, 112, 373, 1960.
71. **Godman, G. C., Morgan, C., Breitenfeld, P. M., and Rose, H. M.**, A correlative study by electron and light microscopy of the development of type 5 adenovirus. II. Light microscopy, *J. Exp. Med.*, 112, 383, 1960.
72. **Weber, J. and Liao, S. K.**, Light and electron microscopy of virus-associated, intranuclear paracrystals in cultured cells infected with types 2, 4, 6, and 18 human *adenoviruses, Can. J. Microbiol.*, 15, 841, 1969.
73. **Henry, C. J., Slifkin, M., Merkow, L. P., and Pardo, M.**, The ultrastructure and nature of adenovirus type 2-induced paracrystalline formations, *Virology*, 44, 215, 1971.
74. **Boulanger, P. A., Torpier, G., and Biserte, G.**, Investigation on intranuclear paracrystalline inclusions induced by Adenovirus 5 in KB cells, *J. Gen. Virol.*, 6, 329, 1970.
75. **Marusyk, R., Norrby, E., and Marusyk, H.**, The relationship of adenovirus-induced paracrystalline structures to the virus core protein(s), *J. Gen. Virol.*, 14, 261, 1972.
76. **Marusyk, R., Marusyk, H., and Lord, T.**, Further evidence on the nature of adenovirus-induced paracrystals, *Microbios*, 6, 71, 1972.
77. **Carstens, E. B. and Marusyk, R. G.**, An image analysis study of adenovirus type 5-induced crystalline inclusions, *J. Gen. Virol.*, 29, 249, 1975.
78. **Boyer, G. S., Denny, F. W., Miller, I., and Ginsberg, H. S.**, Correlation of production of infectious virus with sequential stages of cytologic alteration in HeLa cells infected with adenoviruses types 5 and 7, *J. Exp. Med.*, 112, 865, 1960.
79. **Grayston, J. T., McCarthy, M. A., Smith, M., and Loosli, C. G.**, Adenoviruses. II. The sensitivity of different tissue culture strains of cells to adenoviruses, *J. Infect. Dis.*, 103, 86, 1958.
80. **O'Conor, G. T., Rabson, A. S., Malmgren, R. A., Berezesky, I. K., and Paul, F. K.**, Morphologic observations of green monkey kidney cells after single and double infection with adenovirus 12 and simian virus 40, *J. Natl. Cancer Inst.*, 34, 679, 1965.
81. **Baum, S. G., Horwitz, M. S., and Maizel, J. V.**, Studies of the mechanism of enhancement of human adenovirus infection in monkey cells by simian virus 40, *J. Virol.*, 10, 211, 1972.
82. **Feldman, L. A., Butel, J. S., and Rapp, F.**, Interaction of a simian papovavirus and adenoviruses. I. Induction of adenovirus tumor antigen during abortive infection of simian cells, *J. Bacteriol.*, 91, 813, 1966.
83. **Rabson, A. S., O'Conor, G. T., Berezesky, I. K., and Paul, F. J.**, Enhancement of adenovirus growth in African green monkey kidney cell cultures by SV 40, *Proc. Soc. Exp. Biol. Med.*, 116, 187, 1964.
84. **Goulet, N. R., Schriver, P. W., Liu, O. C., and Beiler, J. M.**,Isolation and propagation of adenoviruses in African green monkey kidney cultures, *J. Infect. Dis.*, 115, 429, 1965.
85. **Klessig, D. F.**, Isolation of a variant of human adenovirus serotype 2 that multiplies efficiently on monkey cells, *J. Virol.*, 21, 1243, 1977.
86. **Klessig, D. F. and Hassell, J. A.**, Characterization of a variant of human adenovirus type 2 which multiplies efficiently in simian cells, *J. Virol.*, 28, 945, 1978.
87. **O'Conor, G. T., Rabson, A. S., Berezesky, I. K., and Paul, F. J.**, Mixed infection with simian virus 40 and adenovirus 12, *J. Natl. Cancer Inst.*, 31, 903, 1963.
88. **Rowe, W. P. and Baum, S. G.**, Evidence for a possible genetic hybrid between adenovirus type 7 and SV 40 viruses, *Proc. Natl. Acad. Sci. U.S.A.*, 52, 1340, 1964.
89. **Rapp, F., Melnick, J. L., Butel, J. S., and Kitahara, T.**, The incorporation of SV 40 genetic material into adenovirus 7 as measured by intranuclear synthesis of SV 40 tumor antigen, *Proc. Natl. Acad. Sci. U.S.A.*, 52, 1348, 1964.

90. **Rowe, W. P. and Baum, S. G.**, Studies of adenovirus SV 40 hybrid viruses. II. Defectiveness of the hybrid particles, *J. Exp. Med.*, 122, 955, 1965.
91. **Boeye, A., Melnick, J. L., and Rapp, F.**, SV 40-adenovirus "Hybrids": presence of two genotypes and the requirement of their complementation for viral replication, *Virology*, 28, 56, 1966.
92. **Rapp, F., Butel, J. S., and Melnick, J. L.**, SV 40-adenovirus "hybrid" populations: transfer of SV 40 determinants from one type of adenovirus to another, *Proc. Natl. Acad. Sci. U.S.A.*, 54, 717, 1965.
93. **Casto, B. C., Armstrong, J. A., Atchison, R. W., and Hammon, W. M.**, Studies on the relationship between adeno-associated virus type 1 (AAV-1) and adenoviruses. II. Inhibition of adenovirus plaques by AAV; its nature and specificity, *Virology*, 33, 452, 1967.
94. **Hoggan, M. D.**, Adenovirus associated viruses, *Prog. Med. Virol.*, 12, 211, 1970.
95. **Malherbe, H. and Harwin, R.**, Seven viruses isolated from the vervet monkey, *Br. J. Exp. Pathol.*, 38, 539, 1957.
96. **Hull, R. N., Johnson, I. S., Culbertson, C. G., Reimer, C. B., and Wright, H. F.**, Oncogenicity of the simian adenoviruses, *Science*, 150, 1044, 1965.
97. **Rapoza, N. P.**, A classification of simian adenoviruses based on hemagglutination, *Am. J. Epidemiol.*, 86, 736, 1967.
98. **Gary, G. W., Hierholzer, J. C., and Black, R. E.**, Characteristics of noncultivable adenoviruses associated with diarrhea in infants: a new subgroup of human adenoviruses, *J. Clin. Microbiol.*, 10, 96, 1979.
99. **Tokumaru, T.**, The nature of toxins of herpes simplex virus. I. Syncytial giant cell producing components in tissue culture, *Arch. Gesamte Virusforsch.*, 24, 104, 1968.
100. **Timbury, M. C., Theriault, A., and Elton, R. A.**, A stable syncytial mutant of herpes simplex type 2 virus, *J. Gen. Virol.*, 23, 219, 1974.
101. **Kousoulas, K. G., Person, S., and Holland, T. C.**, Timing of some of the molecular events required for cell fusion induced by herpes simplex virus type 1, *J. Virol.*, 27, 505, 1978.
102. **Scott, T. F. McN. and McLeod, D. L.**, Cellular responses to infection with strains of herpes simplex virus, *Ann. N. Y. Acad. Sci.*, 81, 118, 1959.
103. **Hoggan, M. D. and Roizman, B.**, The isolation and properties of a variant of herpes simplex producing multinucleated giant cells in monolayer cultures in the presence of antibody, *Am. J. Hyg.*, 70, 208, 1959.
104. **Levitan, D. B. and Blough, H. A.**, Preliminary biochemical characterization of the factor(s) responsible for herpesvirus-induced exogenous fusion, *J. Virol.*, 18, 1081, 1976.
105. **Nii, S. and Kamahora, J.**, Cytopathic changes induced by herpes simplex virus, *Biken J.*, 4, 255, 1961.
106. **Manservigi, R., Spear, P. G., and Buchan, A.**, Cell fusion induced by herpes simplex virus is promoted and suppressed by different viral glycoproteins, *Proc. Natl. Acad. Sci. U.S.A.*, 74, 3913, 1977.
107. **Knowles, R. W. and Person, S.**, Effects of 2-Deoxyglucose, glucosamine, and mannose on cell fusion and the glycoproteins of herpes simplex virus, *J. Virol.*, 18, 644, 1976.
108. **Bungay, C., and Watkins, J. F.**, Observations on polykaryocytosis in HeLa cells infected with herpes simplex virus, *Br. J. Exp. Pathol.*, 45, 48, 1964.
109. **Love, R. and Wildy, P.**, Cytochemical studies of the nucleoproteins of HeLa cells infected with herpes virus, *J. Cell Biol.*, 17, 237, 1963.
110. **Crouse, H. V., Coriell, L. L., Blank, H., and Scott, T. F.**, Cytochemical studies on the intranuclear inclusion of herpes simplex, *J. Immunol.*, 65, 119, 1950.
111. **Kristensson, K., Enerback, L., and Sourander, P.**, Histochemical and ultrastructural properties of the classical inclusion body in neurones infected with herpes simplex virus, *Acta Pathol. Microbiol. Scand. Sect. A*, 78, 595, 1970.
112. **Morgan, C., Ellison, S. A., Rose, H. M., and Moore, D. H.**, Structure and development of viruses as observed in the electron microscope. I. Herpes simplex virus, *J. Exp. Med.*, 100, 195, 1954.
113. **Morgan, C., Rose, H. M., Holden, M., and Jones, E. P.**, Electron microscopic observations on the development of herpes simplex virus, *J. Exp. Med.*, 110, 643, 1959.
114. **Duff, R. and Rapp, F.**, Oncogenic transformation of hamster cells after exposure to herpes simplex virus type 2, *Nature (London) New Biol.*, 233, 48, 1971.
115. **Sabin, A. B. and Wright, A. M.**, Acute ascending myelitis following a monkey bite, with the isolation of a virus capable of reproducing the disease, *J. Exp. Med.*, 59, 115, 1934.
116. **Orcutt, R. P., Pucak, G. J., Foster, H. L., Kilcourse, J. T., and Ferrell, T.**, Multiple testing for the detection of B virus antibody in specially handled rhesus monkeys after capture from virgin trapping grounds, *Lab. Anim. Sci.*, 26, 70, 1976.
117. Center for Disease Control, U.S.A., Herpes B encephalitis, *Morbid. Mortal. Wkly. Rep.*, 22, 333, 1973.
118. **Reissig, M., and Melnick, J. L.**, The cellular changes produced in tissue cultures by Herpes B virus correlated with the concurrent multiplication of the virus, *J. Exp. Med.*, 101, 341, 1955.

119. **Benda, R. and Cinatl, J.**, Isolation of two plaque variants from the prototype strain of B virus (*Herpesvirus simiae*), *Acta Virol. Engl. Ed.*, 10, 178, 1966.
120. **Falke, D.**, Isolation of two variants with different cytopathic properties from a strain of herpes B virus, *Virology*, 14, 492, 1961.
121. **Hull, R. N.**, The simian viruses, *Virol. Monogr.*, 2, 1, 1968.
122. **Malherbe, H. and Harwin, R.**, Neurotropic virus in African monkeys, *Lancet*, 2, 530, 1958.
123. **Malherbe, H. and Strickland-Cholmley, M.**, Simian herpesvirus SA 8 from a baboon, *Lancet*, 2, 1427, 1969.
124. **Malherbe, H. and Strickland-Cholmley, M.**, Virus from baboons, *Lancet*, 2, 1300, 1969.
125. **Nesburn, A. B.**, Isolation and characterization of a herpeslike virus from New Zealand albino rabbit kidney cell cultures: a probable reisolation of Virus 111 of Rivers, *J. Virol.*, 3, 59, 1969.
126. **Malherbe, H. and Strickland-Cholmley, M.**, Herpesvirus latent in primary rabbit kidney cell cultures, *Lancet*, 2, 930, 1972.
127. **Waner, J. L. and Weller, T. H.**, Serological and cultural studies bearing on the persistent nature of cytomegaloviral infections in man, *Perspect. Virol.*, 8, 211, 1973.
128. **Waner, J. L. and Weller, T. H.**, Behavior of human cytomegaloviruses in cell cultures of bovine and simian origin, *Proc. Soc. Exp. Biol. Med.*, 145, 379, 1974.
129. **Diosi, P., Babusceac, L., and Gherman, D.**, Cytophagia in cell cultures infected with cytomegalovirus, *J. Infect. Dis.*, 125, 669, 1972.
130. **Diosi, P., Babusceac, L., and David, C.**, Distinctive cytopathic properties of cytomegalovirus recovered from the female genital tract, *Am. J. Clin. Pathol.*, 59, 192, 1973.
131. **Black, P. H., Hartley, J. W., and Rowe, W. P.**, Isolation of a cytomegalovirus from African green monkey, *Proc. Soc. Exp. Biol. Med.*, 112, 601, 1963.
132. **Knowles, W. A.**, In vitro cultivation of human cytomegalovirus in thyroid epithelial cells, *Arch. Virol.*, 50, 119, 1976.
133. **Vonka, V., Anisimova, E., and Macek, M.**, Replication of cytomegalovirus in human epitheloid diploid cell line, *Arch. Virol.*, 52, 283, 1976.
134. **Figueroa, M. E., Geder, L., and Rapp, F.**, Infection of human amnion cells with cytomegalovirus, *J. Med. Virol.*, 2, 369, 1978.
135. **Vesterinen, E., Leinikki, P., and Saksela, E.**, Cytopathogenicity of cytomegalovirus to human ecto- and endocervical epithelial cells in vitro, *Acta Cytol.*, 19, 473, 1975.
136. **Jeor, St., S. and Rapp, F.**, Cytomegalovirus: conversion of nonpermissive cells to a permissive state for virus replication, *Science*, 181, 1060, 1973.
137. **Michelson-Fiske, S., Arnoult, J., and Febvre, H.**, Cytomegalovirus infection of human lung epithelial cells in vitro, *Intervirology*, 5, 354, 1975.
138. **Smith, A. J. and Weidman, F. D.**, Further note upon the occurrence of Endamoeba mortinatalium as a human parasite, *Am. J. Trop. Dis.*, 2, 256, 1914.
139. **Goodpasture, E. W. and Talbot, F. B.**, Concerning the nature of "protozoan-like" cells in certain lesions of infancy, *Am. J. Dis. Child.*, 21, 415, 1921.
140. **Weller, T. H., Hanshaw, J. B., and Scott, D. E.** Serologic differentiation of viruses responsible for cytomegalic inclusion disease, *Virology*, 12, 130, 1960.
141. **Weller, T. H.**, Serial propagation in vitro of agents producing inclusion bodies derived from varicella and herpes zoster, *Proc. Soc. Exp. Biol. Med.*, 83, 340, 1953.
142. **Weller, T. H., Witton, H. M., and Bell, E. J.**, The etiologic agents of varicella and herpes zoster. Isolation, propagation, and cultural characteristics in vitro, *J. Exp. Med.*, 108, 843, 1958.
143. **Weller, T. H.**, Varicella-Herpes zoster virus, in *Viral and Rickettsial Infections of Man*, 4th ed., Horsfall, F. L. and Tamm, I., Eds., Lippincott, Philadelphia, 1965, chap. 42.
144. **Weller, T. H.**, Varicella-zoster virus, in *Diagnostic Procedures for Viral and Rickettsial Infections*, 4th ed., Lennette, E. H. and Schmidt, N. J., Eds., American Public Health Association, New York, 1969, chap. 21.
145. **Rapp, F. and Benyesh-Melnick, M.**, Plaque assay for measurement of cells infected with zoster virus, *Science*, 141, 433, 1963.
146. **Grose, C. and Brunell, P. A.**, Varicella-zoster virus: isolation and propagation in human melanoma cells at 36 and 32°C, *Infect. Immun.*, 19, 199, 1978.
147. **Darai, G. and Flugel, R. M.**, Mink lung cells: a non-primate cell line highly susceptible for varicella-zoster virus, *J. Gen. Virol.*, 41, 179, 1978.
148. **Taylor-Robinson, D. and Caunt, A. E.**, Varicella virus, *Virol. Monogr.*, 12, 1, 1972.
149. **Taylor-Robinson, D.**, Chickenpox and herpes zoster. III. Tissue culture studies, *Br. J. Exp. Pathol.*, 40, 521, 1959.
150. **Caunt, A. E. and Taylor-Robinson, D.**, Cell-free varicella-zoster virus in tissue culture, *J. Hyg.*, 62, 413, 1964.
151. **Clarkson, M. J., Thorpe, E., and McCarthy, K.**, A virus disease of captive vervet monkeys (Cercopithecus aethiops) caused by a new herpesvirus, *Arch. Gesamte Virusforsch.*, 22, 219, 1967.

152. **McCarthy, K., Thorpe, E., Laursen, A. C., Heymann, C. S., and Beale, A. J.,** Exanthematous disease in patas monkeys caused by a herpes virus, *Lancet*, 2, 856, 1968.
153. **Ayres, J. P.,** Studies of the Delta herpesvirus isolated from the patas monkey (Erythrocebus patas), *Lab. Anim. Sci.*, 21, 685, 1971.
154. **Blakeley, G. A., Lourie, B., Morton, W. G., Evans, H. H., and Kaufmann, A. F.,** A varicella-like disease in macaque monkeys, *J. Infect. Dis.*, 127, 617, 1973.
155. **Felsenfeld, A. D. and Schmidt, N. J.,** Immunological relationship between Delta herpesvirus of patas monkeys and varicella-zoster virus of humans, *Infect. Immun.*, 12, 261, 1975.
156. **Felsenfeld, A. D. and Schmidt, N. J.,** Antigenic relationships among several simian varicella-like viruses and varicella-zoster virus, *Infect. Immun.*, 15, 807, 1977.

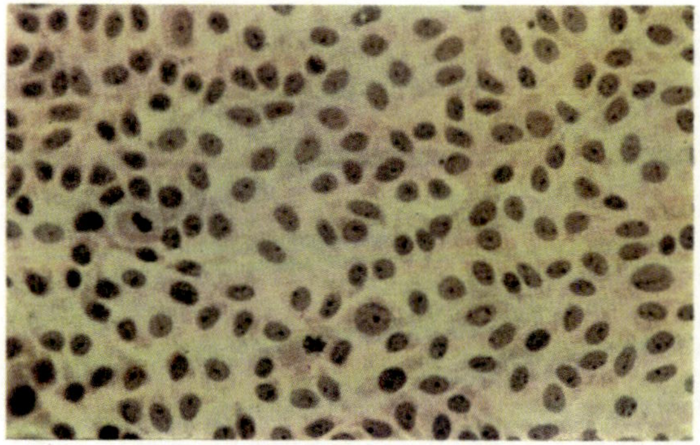

FIGURE 1. Uninfected vervet monkey kidney cells. (H & E; magnification × 200.)

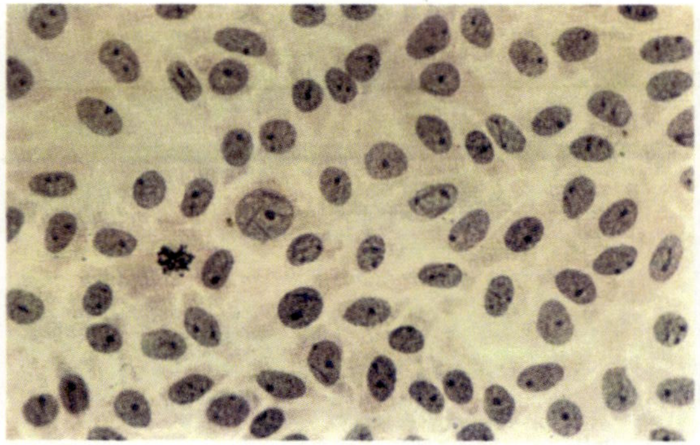

FIGURE 2. Uninfected vervet monkey kidney cells. (H & E; magnification × 320.)

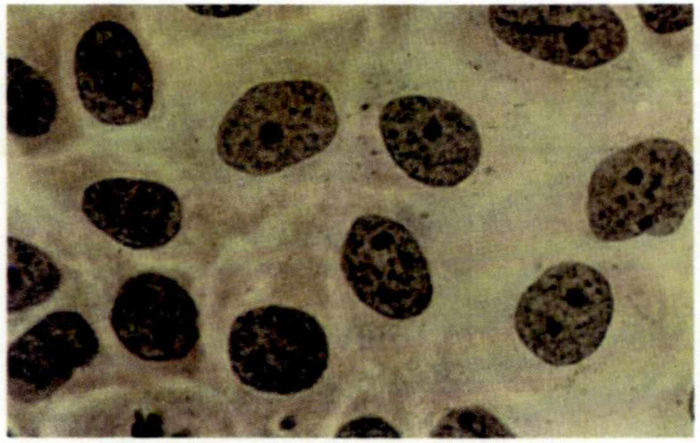

FIGURE 3. Uninfected vervet monkey kidney cells. (H & E; magnification × 800.)

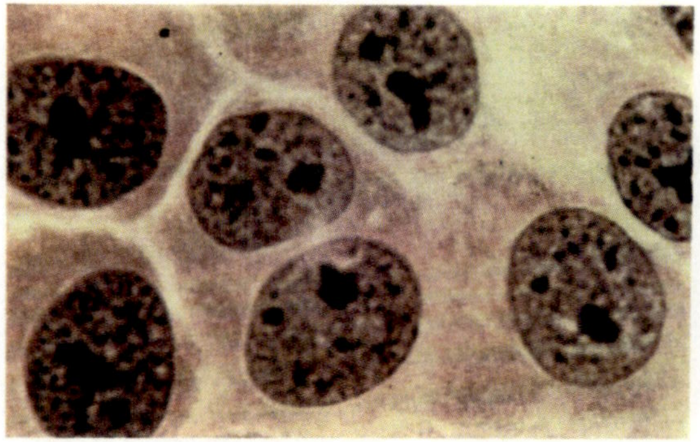

FIGURE 4. Uninfected HEp-2 tumor cell line. (H & E; magnification × 1250.)

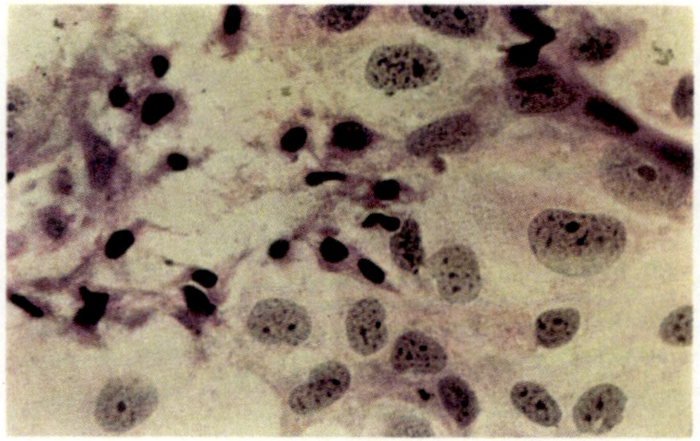

FIGURE 5. Nonviral (? bacterial) toxic changes. Vervet kidney cells. (H & E; magnification × 320.)

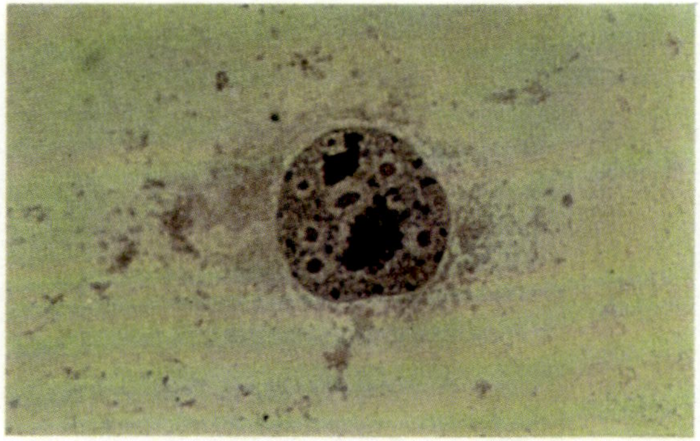

FIGURE 6. Amoebic toxin. Eosinophilic bodies in kidney cell nucleus. (H & E; magnification × 800.)

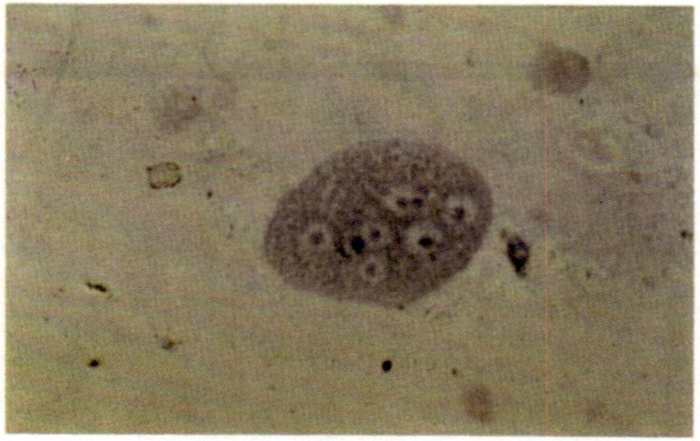

FIGURE 7. Actinomycin D. Eosinophilic bodies in kidney cell nucleus. (H & E; magnification × 800.)

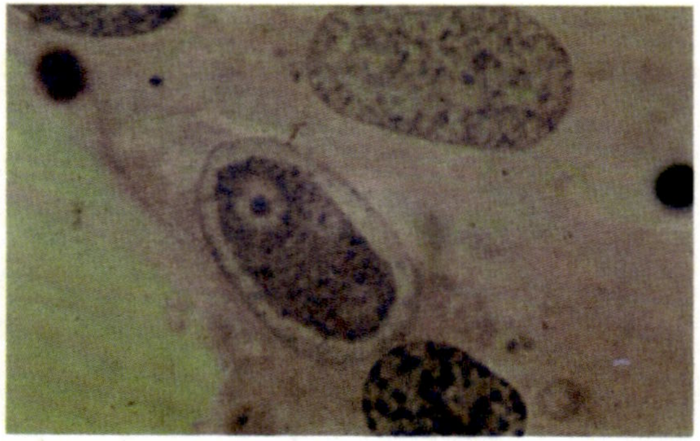

FIGURE 8. Actinomycin D. Basophilic retracted body in kidney cell nucleus. (H & E; magnification × 1000.)

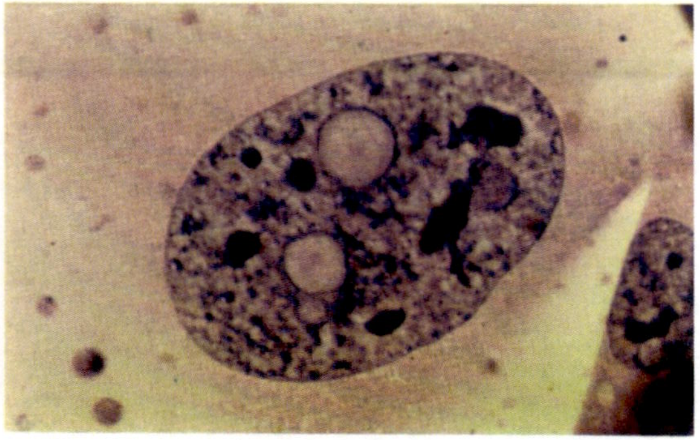

FIGURE 9. Nuclear vacuoles. Uninfected embryo monkey kidney cell line. (H & E; magnification × 800.)

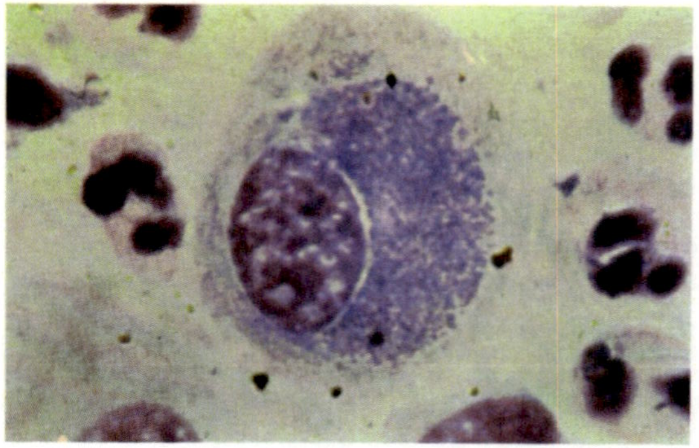

FIGURE 10. Trachoma (Halberstaedter-Prowazek) body. Conjunctival epithelial cell. (Giemsa; magnification × 800.)

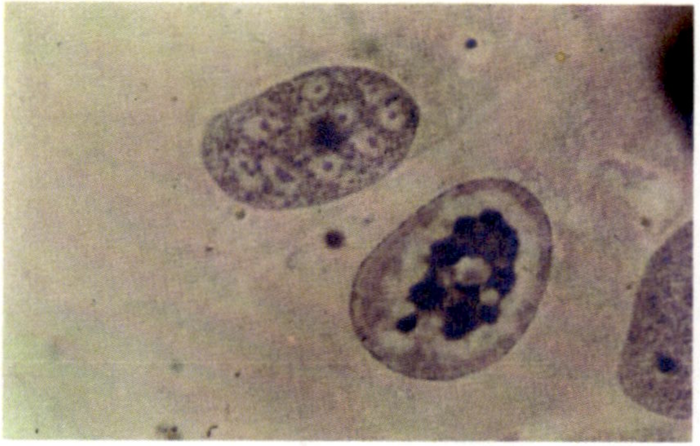

FIGURE 11. Early (type B) and late (type A) adenovirus inclusions. Vervet kidney cells. (H & E; magnification × 800.)

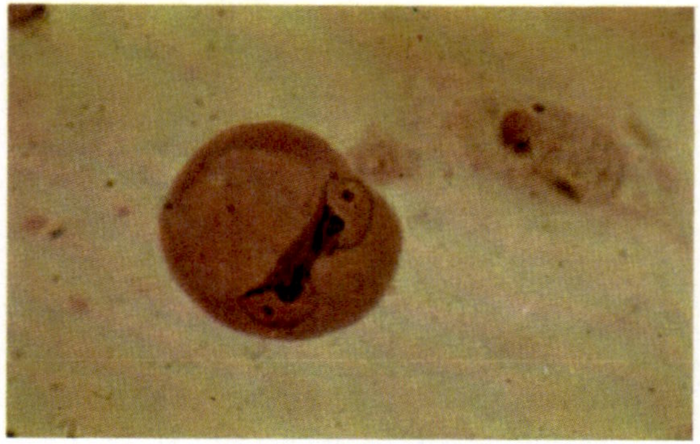

FIGURE 12. Picornavirus paranuclear zone. Simian enterovirus SA 4 in vervet kidney cells. (H & E; magnification × 800.)

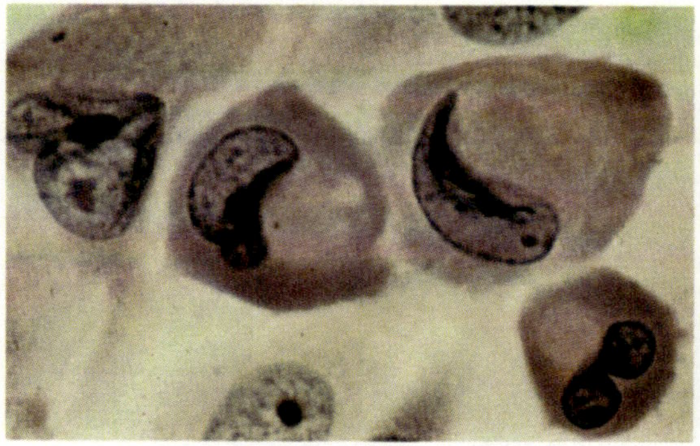

FIGURE 13. Picornavirus paranuclear zones. Simian enterovirus SA 5 in vervet kidney cells. (H & E; magnification × 800.)

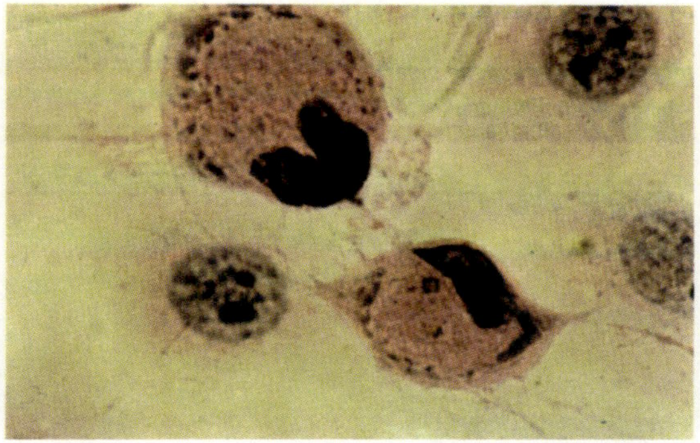

FIGURE 14. Picornavirus basophilic cytoplasmic bodies. Simian enterovirus SV 6 in baboon kidney cells. (H & E; magnification × 800.)

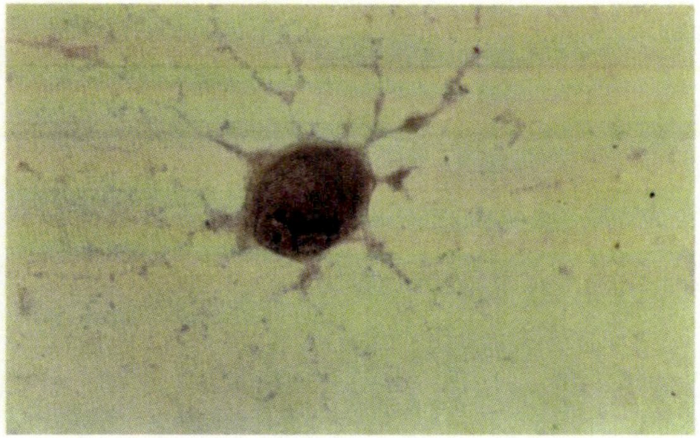

FIGURE 15. Picornavirus coxsackie B 3. Cytoplasmic processes. MA 104 monkey kidney line. (H & E; magnification × 320.)

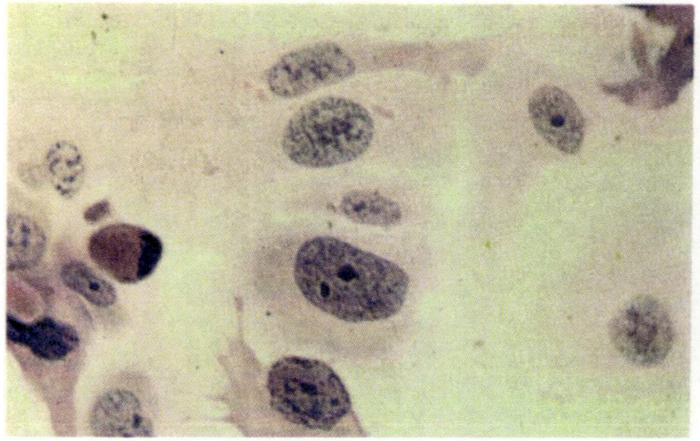

FIGURE 16. Nucleolar dispersion. Simian enterovirus SV 44 in baboon kidney cells. (H & E; magnification × 320.)

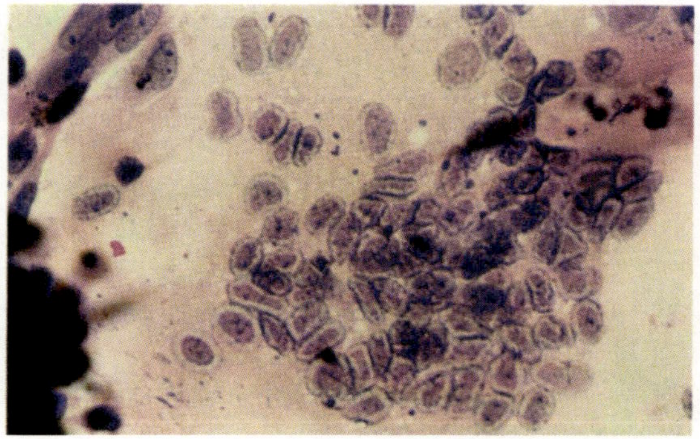

FIGURE 17. Poliovirus and spumavirus together in vervet kidney cells. (H & E; magnification × 320.)

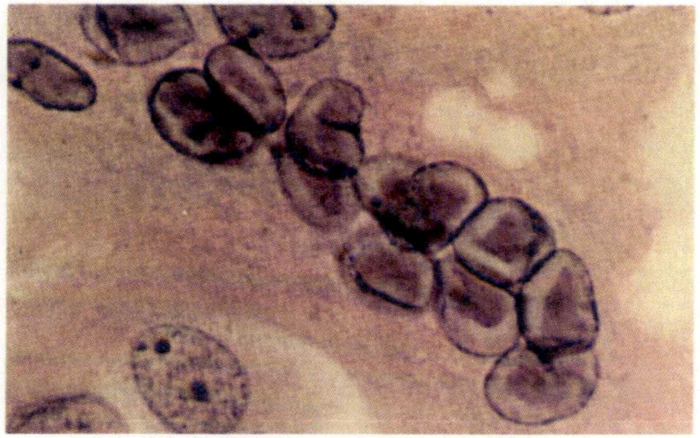

FIGURE 18. Echovirus type 1 and spumavirus together in vervet kidney cells. (H & E; magnification × 800.)

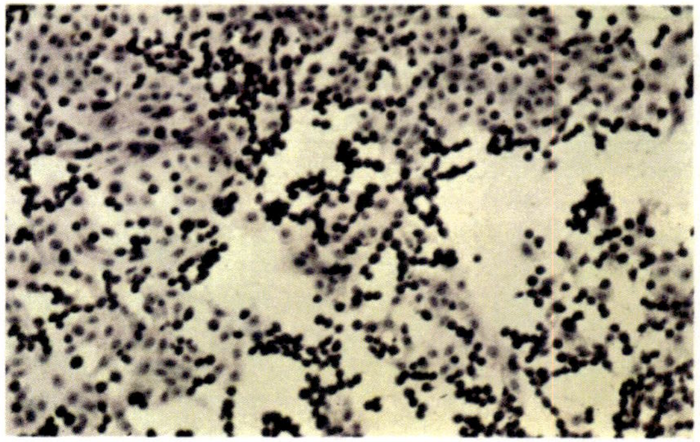

FIGURE 19. CPE at low magnification. Coxsackievirus B 2 in Vero cells. (H & E; magnification × 80.)

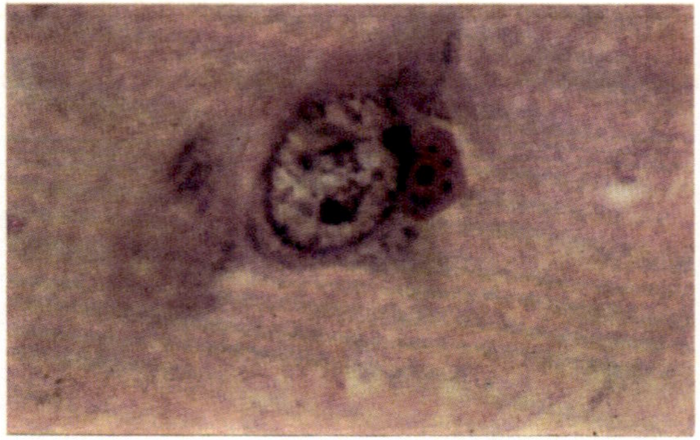

FIGURE 20. Negri body in cytoplasm of neurone. Mouse brain. (H & E; magnification × 1000.)

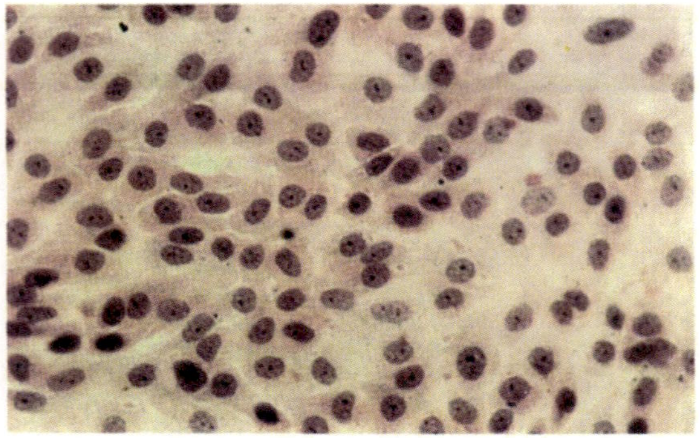

FIGURE 21. Marburg virus, tenth day in vervet monkey kidney cells. Sheet intact. (H & E; magnification × 200.)

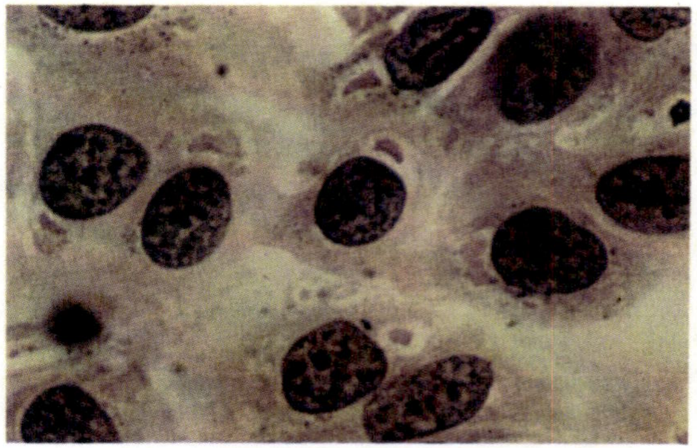

FIGURE 22. Marburg virus, same culture as in Figure 21. Cytoplasmic inclusions. (Magnification × 800.)

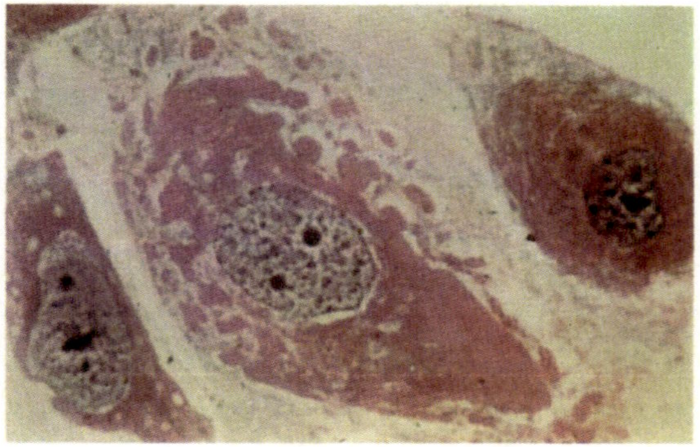

FIGURE 23. Reovirus in vervet monkey kidney cells. (H & E; magnification × 1250.)

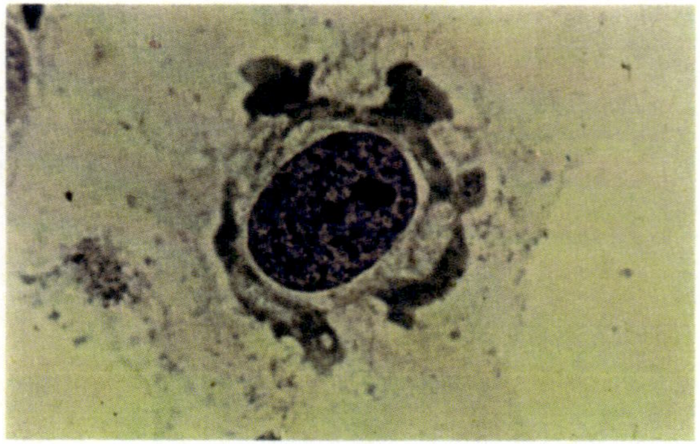

FIGURE 24. Reovirus in MA 104 rhesus kidney cell line. (Giemsa; magnification × 800.)

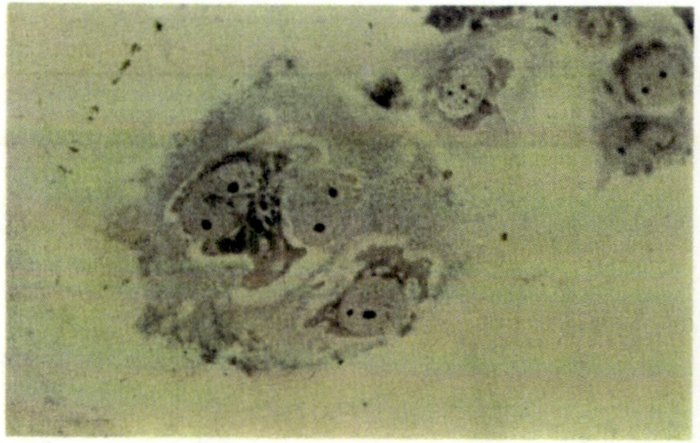

FIGURE 25. Reovirus in MA 104 rhesus kidney cell line. (May-Grünwald-Giemsa; magnification × 320.)

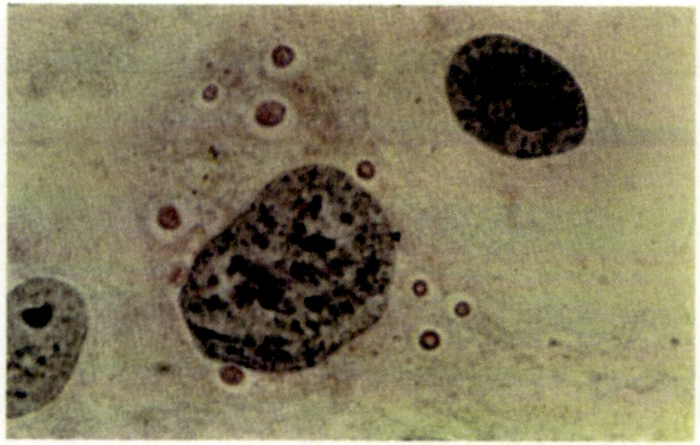

FIGURE 26. Simian rotavirus SA 11 in MA 104 rhesus kidney cell line. (H & E; magnification × 800.)

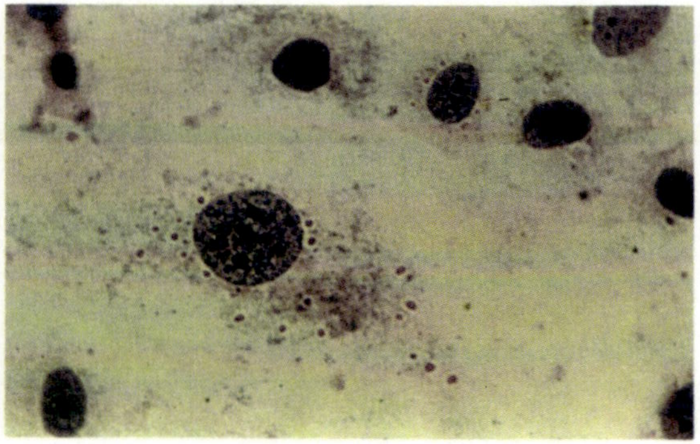

FIGURE 27. Nebraska calf scours rotavirus in MA 104 rhesus kidney cell line. (H & E; magnification × 320.)

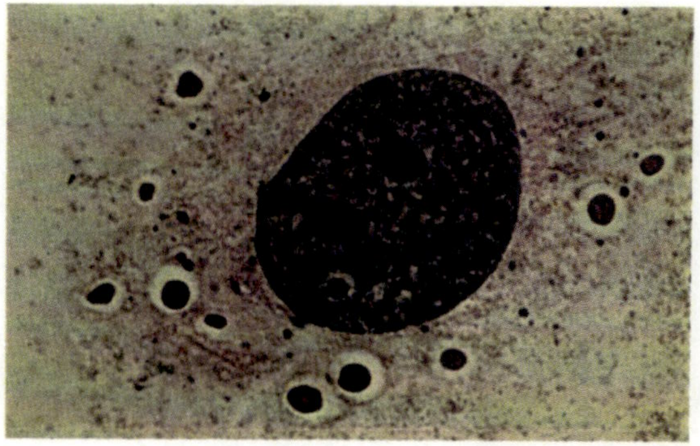

FIGURE 28. Nebraska calf scours rotavirus in MA 104 rhesus kidney cell line. (H & E; magnification × 800.)

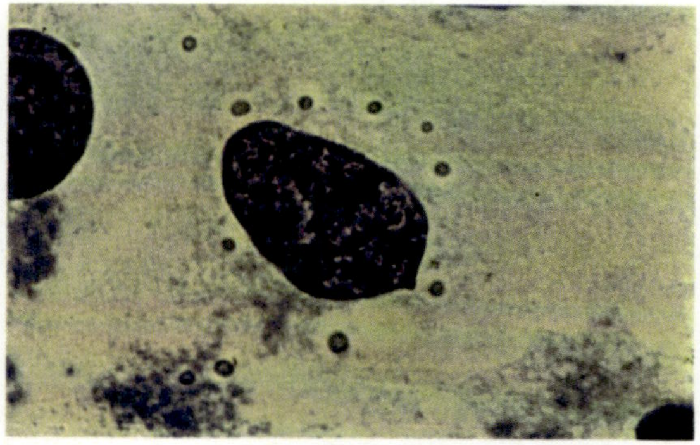

FIGURE 29. Simian rotavirus SA 11 in MA 104 rhesus kidney cell line. (Giemsa; magnification × 800.)

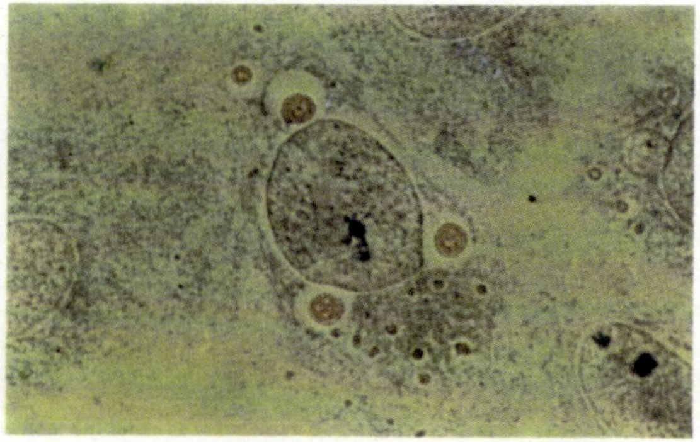

FIGURE 30. Simian rotavirus SA 11 in MA 104 rhesus kidney cell line. (May-Grunwald-Giemsa; magnification × 800.)

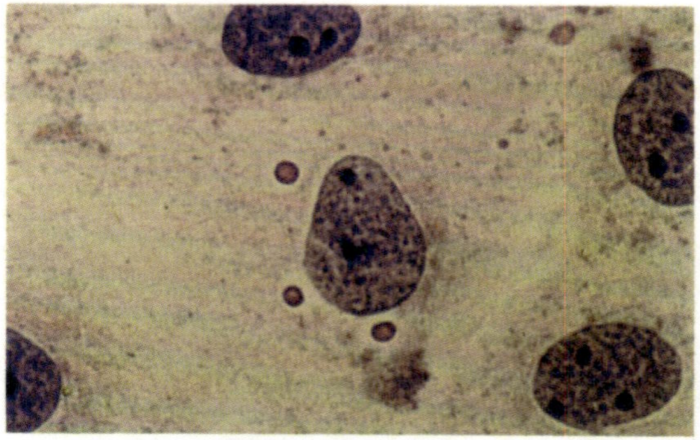

FIGURE 31. Simian rotavirus SA 11 in MA 104 rhesus kidney cell line. (H & E; magnification × 800.)

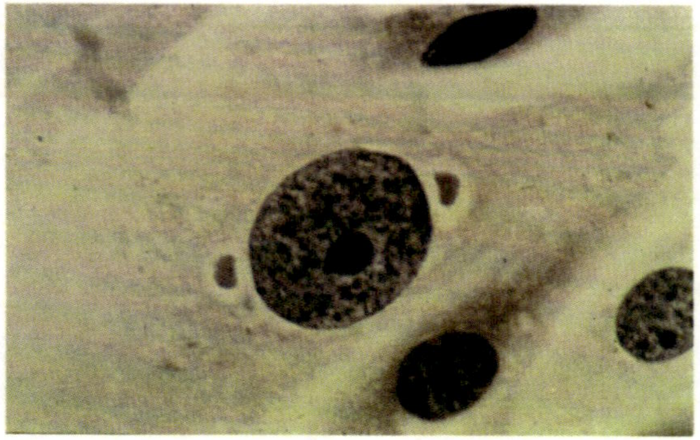

FIGURE 32. Bluetongue virus in Vero cells. Day 2. (H & E; magnification × 800.)

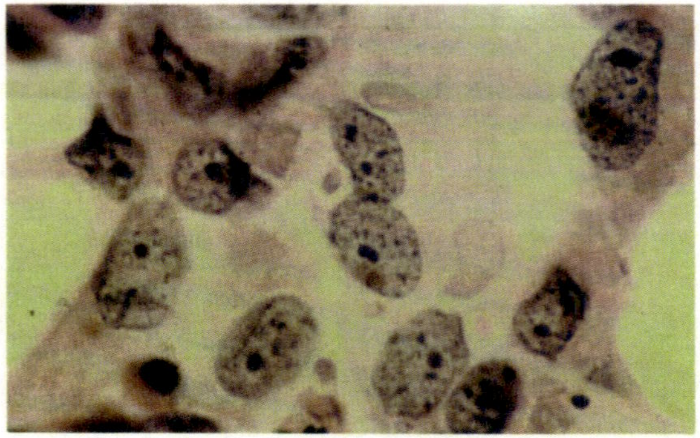

FIGURE 33. Variola minor virus in vervet kidney cells. Eosinophilic and basophilic cytoplasmic inclusions. (H & E; magnification × 800.)

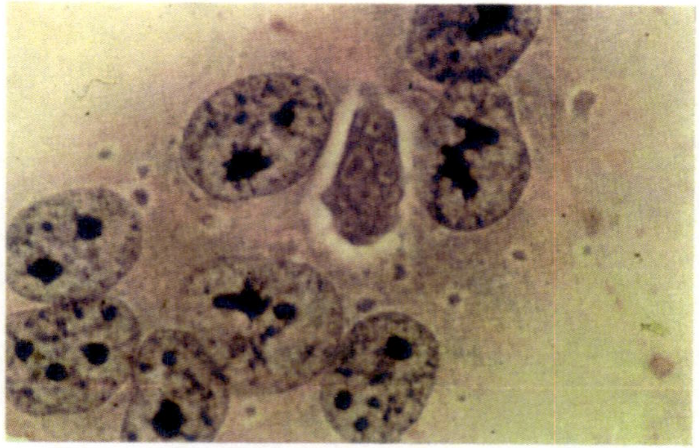

FIGURE 34. Variola minor virus in embryo monkey kidney cells. Basophilic inclusion with internal structures. (H & E; magnification × 800.)

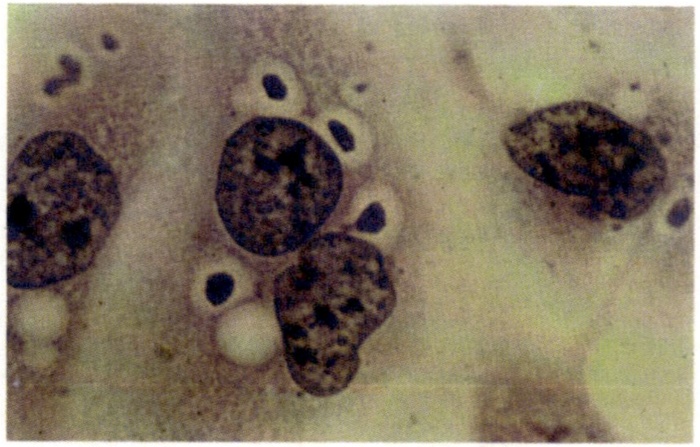

FIGURE 35. Variola minor virus in vervet monkey kidney cells. Basophilic cytoplasmic inclusions. (H & E; magnification × 800.)

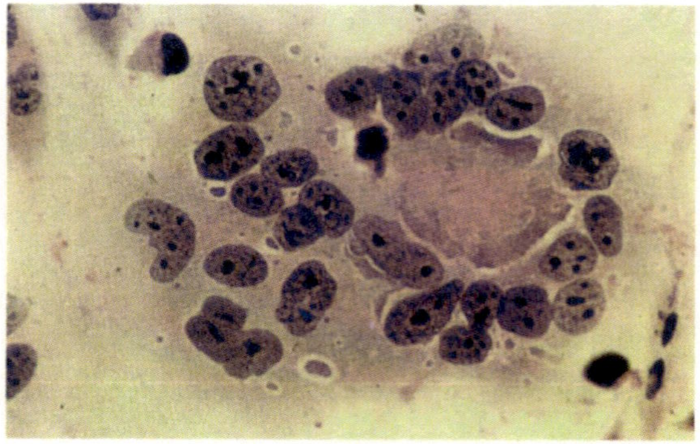

FIGURE 36. Variola minor virus in embryo monkey kidney cells. Syncytium with basophilic inclusions. (H & E; magnification × 320.)

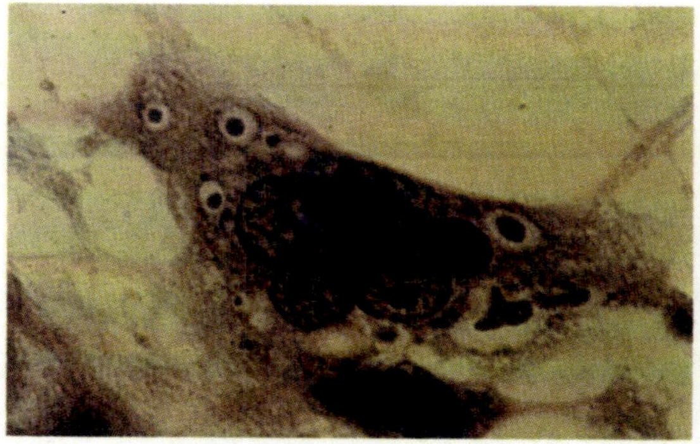

FIGURE 37. Vaccinia virus in Vero cells. Basophilic inclusions. (H & E; magnification × 800.)

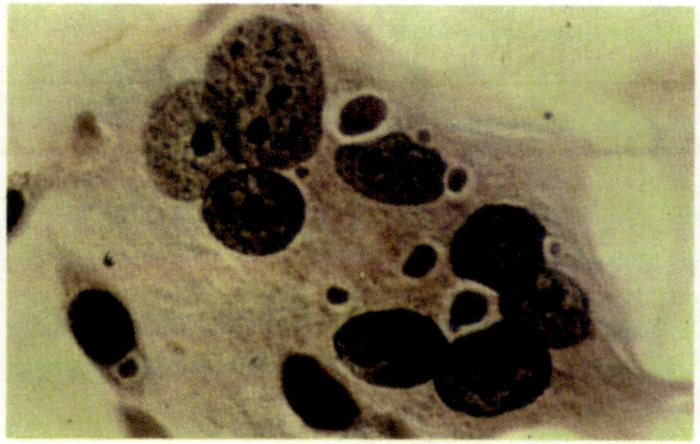

FIGURE 38. Monkeypox virus in Vero cells. Basophilic inclusions. (H & E; magnification × 800.)

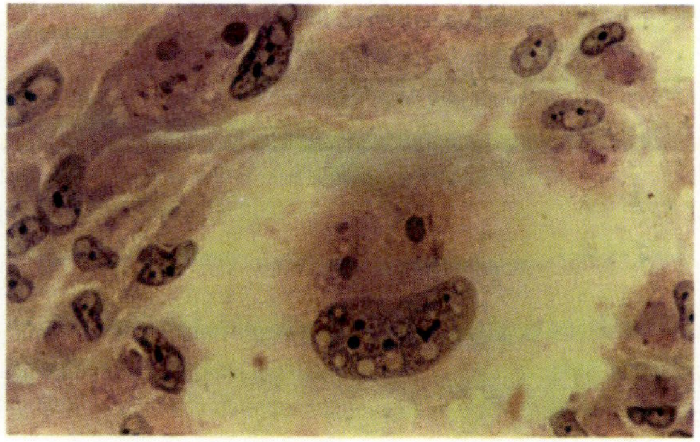

FIGURE 39. Tanapox virus in human embryo kidney cells. (H & E; magnification × 320.)

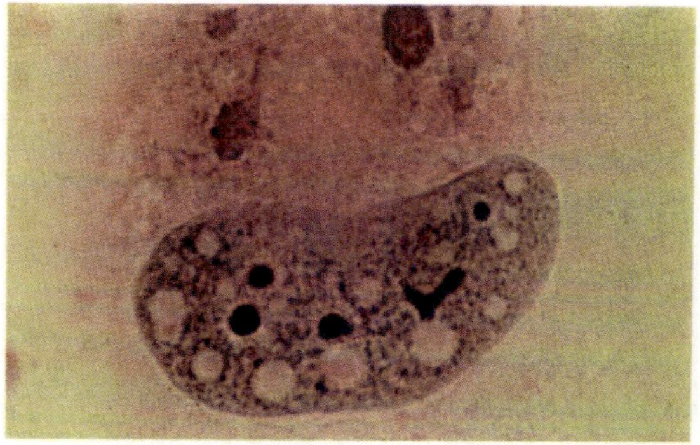

FIGURE 40. Tanapox virus in human embryo kidney cell. (H & E; magnification × 800.)

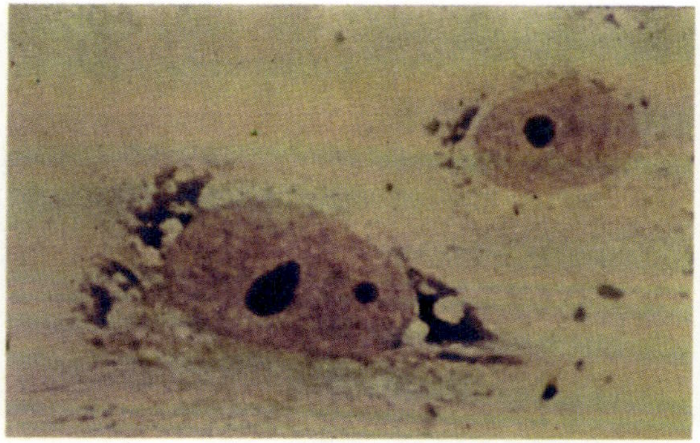

FIGURE 41. Lassa virus in Vero cells. (Giemsa; magnification × 800.)

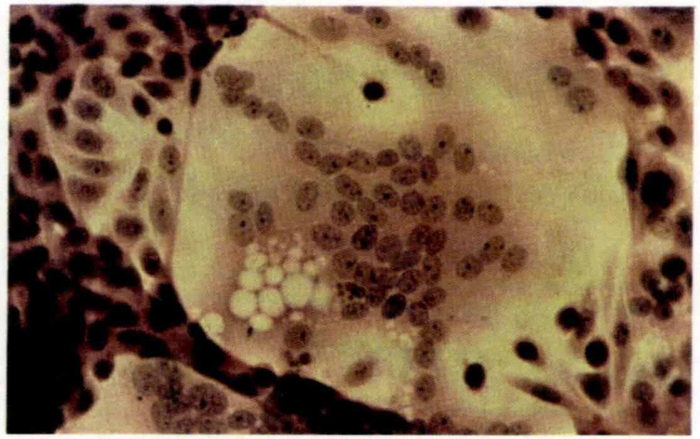

FIGURE 42. Simian spumavirus in vervet monkey kidney cells. Slight vacuolation. (H & E; magnification × 200.)

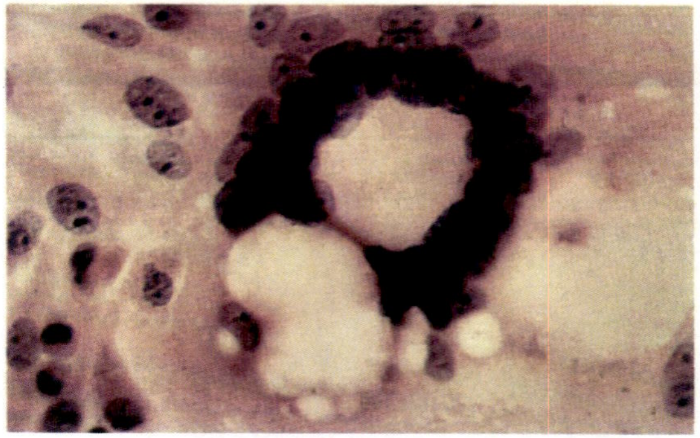

FIGURE 43. Simian spumavirus in vervet monkey kidney cells. Foamy effect. (H & E; magnification × 320.)

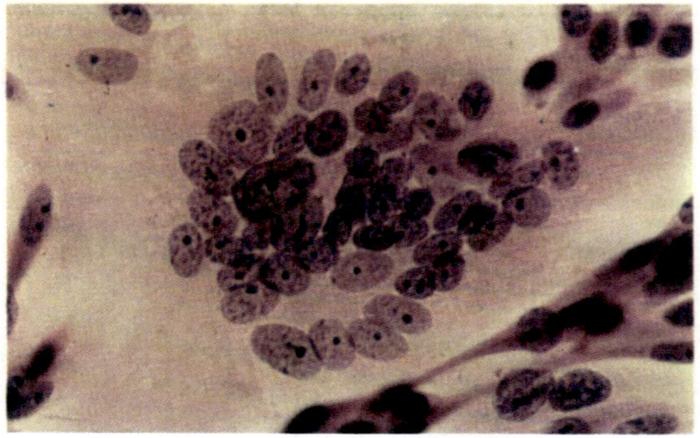

FIGURE 44. Simian spumavirus in vervet monkey kidney cells. No vacuolation. (H & E; magnification × 320.)

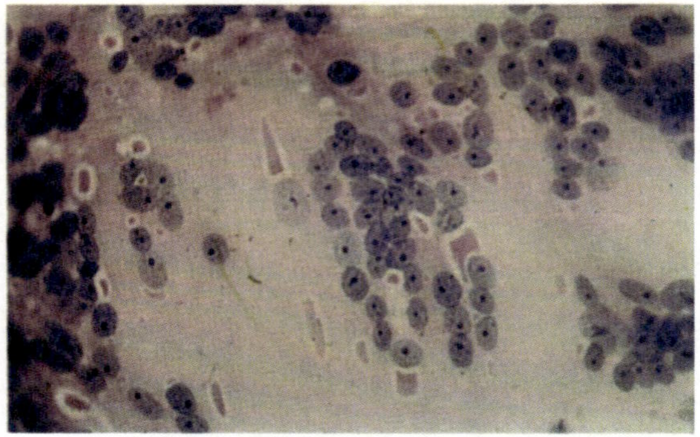

FIGURE 45. Respiratory syncytial virus in Vero cells. (H & E; magnification × 200.)

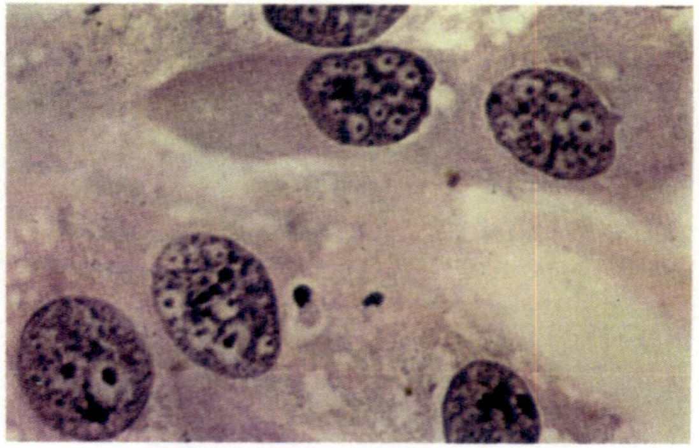

FIGURE 46. Sindbis virus in vervet monkey kidney cells. (H & E; magnification × 800.)

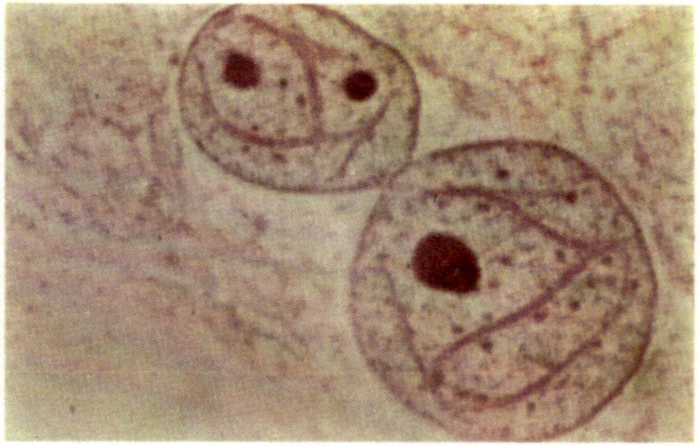

FIGURE 47. Rift Valley Fever virus in vervet monkey kidney cells. Filamentous nuclear inclusions. (H & E; magnification × 1250.)

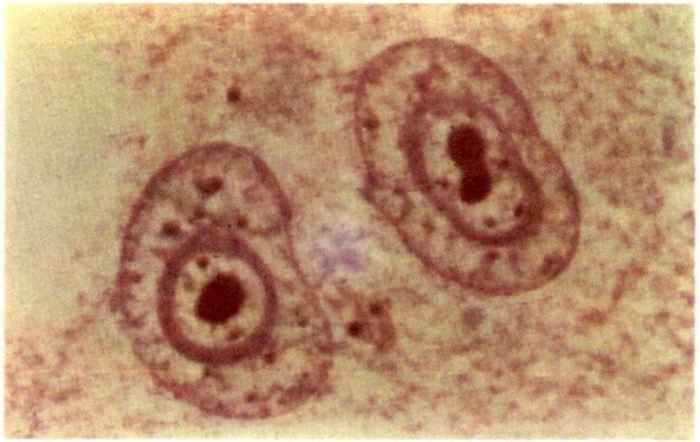

FIGURE 48. Rift Valley Fever virus in vervet monkey kidney cells. Annular nuclear inclusions. (H & E; magnification × 1250.)

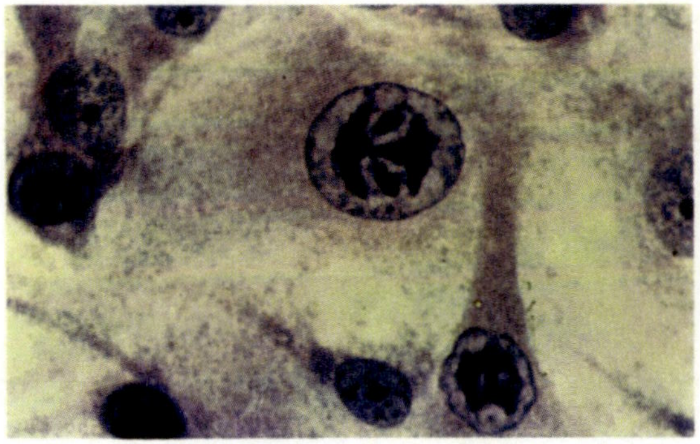

FIGURE 49. Feline panleukopenia virus (parvovirus) in CRFK feline kidney cells. (H & E; magnification × 800.)

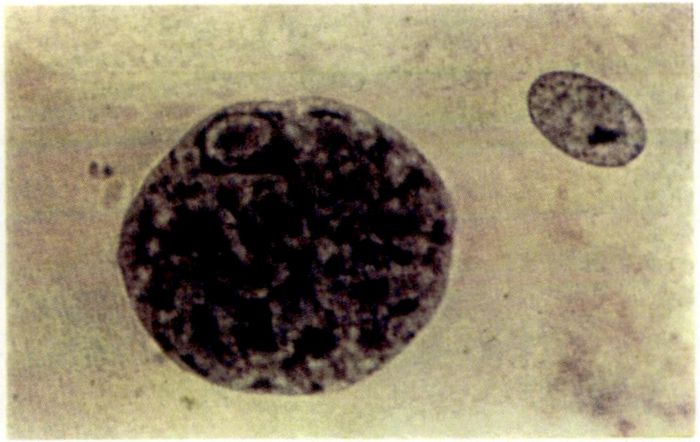

FIGURE 50. SV 40 in baboon kidney cell. (H & E; magnification × 800.)

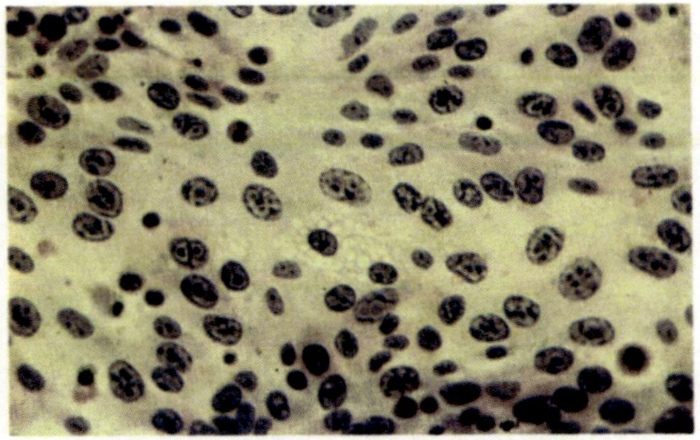

FIGURE 51. SV 40 in baboon kidney cells. Many nuclear inclusions, scanty vacuolation. (H & E; magnification × 200.)

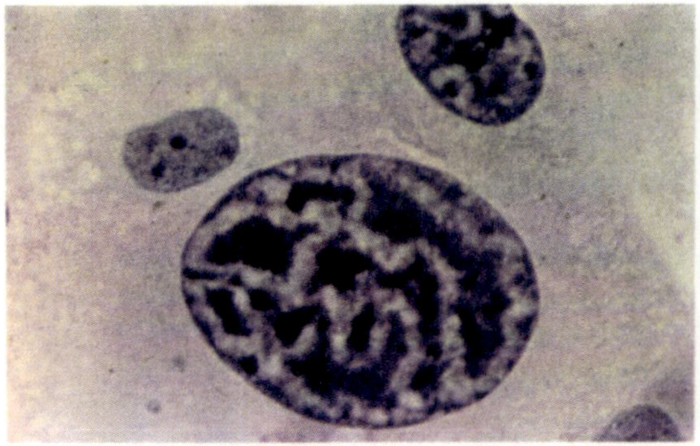

FIGURE 52. Simian papovavirus SA 12 in vervet monkey kidney cells. (H & E; magnification × 800.)

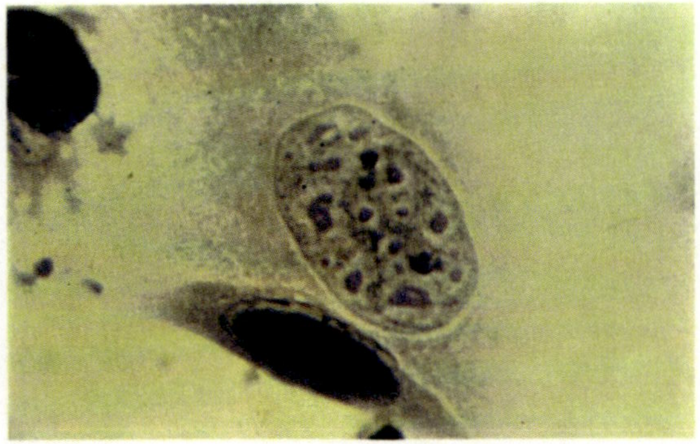

FIGURE 53. Human adenovirus in vervet monkey kidney cells. Early multiple nuclear inclusions. (H & E; magnification × 800.)

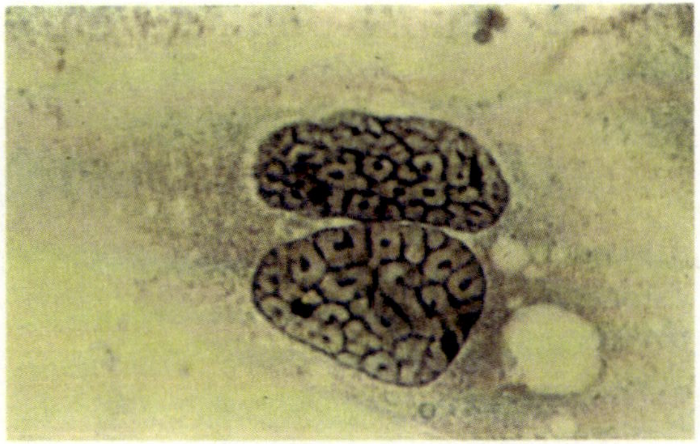

FIGURE 54. Simian adenovirus SA 18 in baboon kidney cells. Early discrete inclusions. (H & E; magnification × 800.)

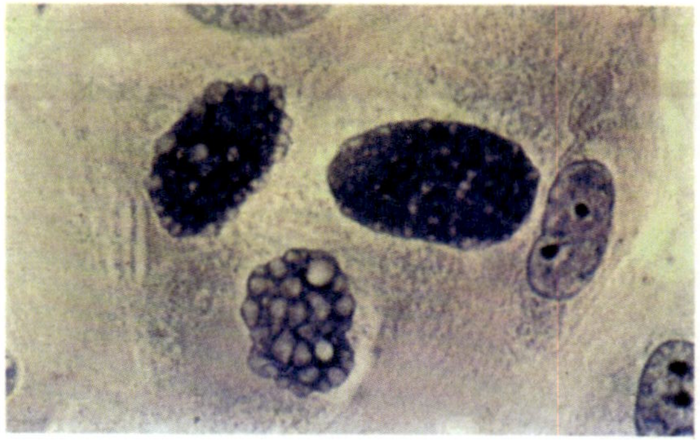

FIGURE 55. Simian adenovirus strain in baboon kidney cells. Moruloid forms. (H & E; magnification × 800.)

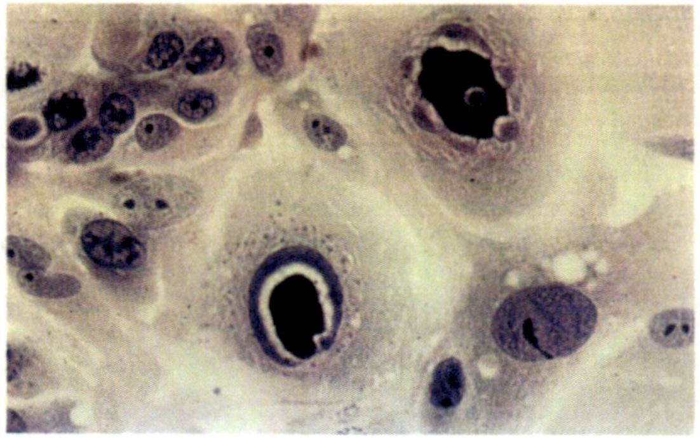

FIGURE 56. Simian adenovirus SV 30 in baboon kidney cells. Nuclear bubbles, and late stage with cytoplasmic inclusions. (H & E; magnification × 320.)

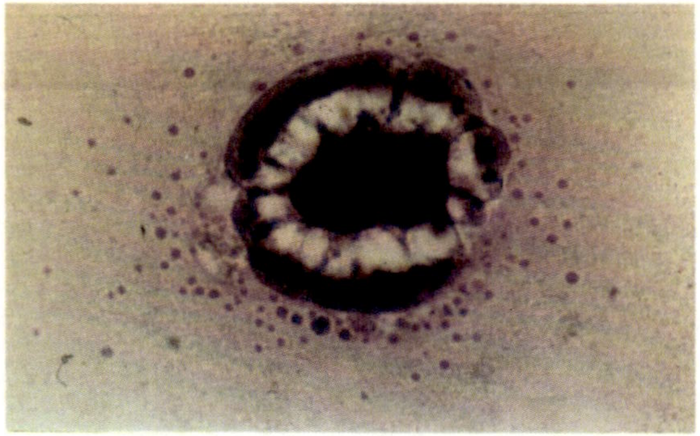

FIGURE 57. Simian adenovirus strain in baboon kidney cell. Late nuclear stage with cytoplasmic inclusions. (H & E; magnification × 800.)

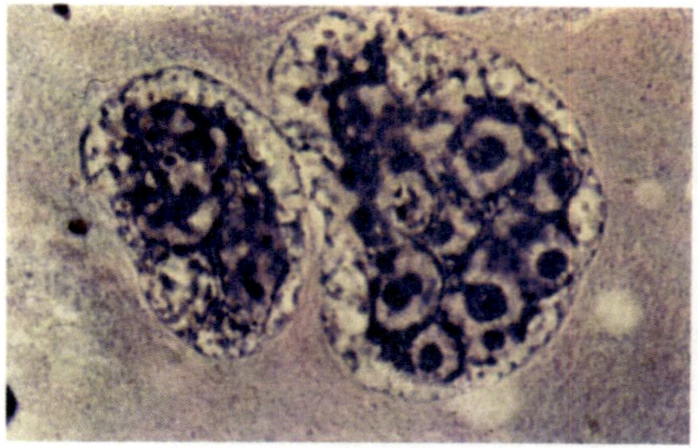

FIGURE 58. Simian adenovirus SV 20 in baboon kidney cells. Intermediate nuclear stage with basophilic structures. (H & E; magnification × 800.)

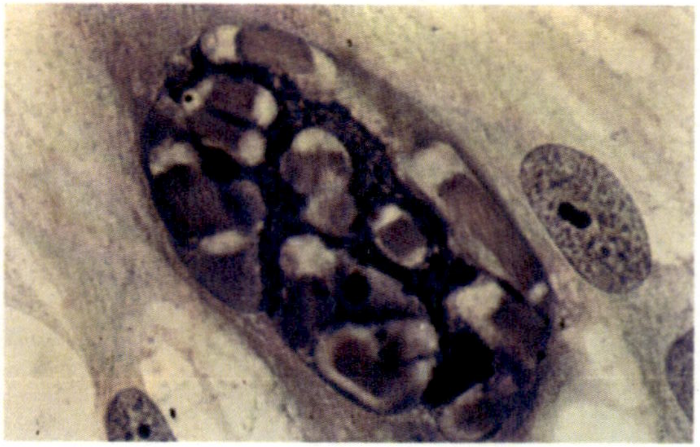

FIGURE 59. Human adenovirus type 5 in human diploid lung cell. Nuclear bubbles with eosinophilic bars. (H & E; magnification × 800.)

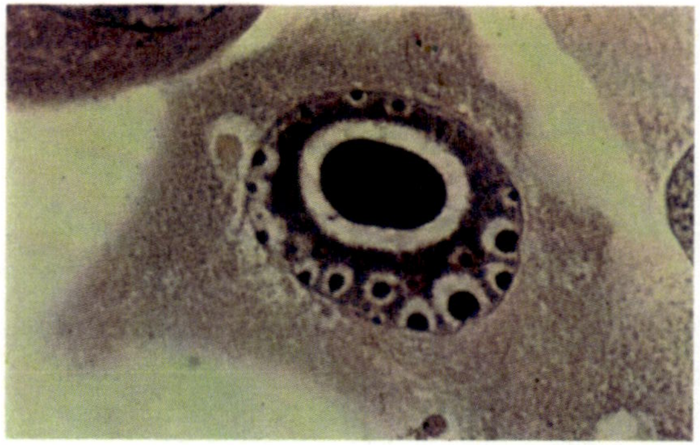

FIGURE 60. Simian adenovirus strain in baboon kidney cell. Late nuclear stage with basophilic bodies. (H & E; magnification × 800.)

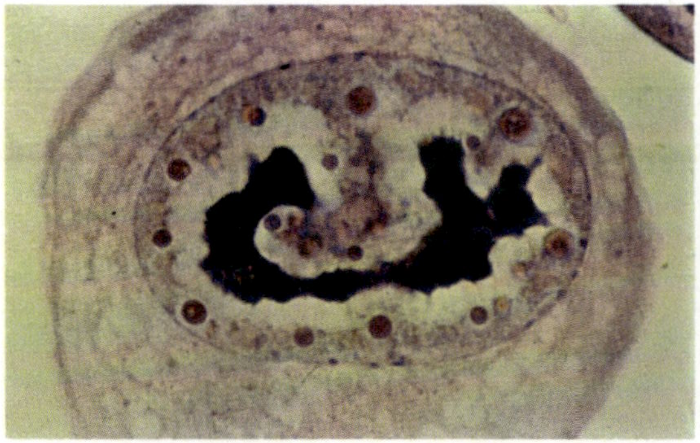

FIGURE 61. Simian adenovirus strain in baboon kidney cell. Late stage, with red bodies in nucleus. (H & E; magnification × 800.)

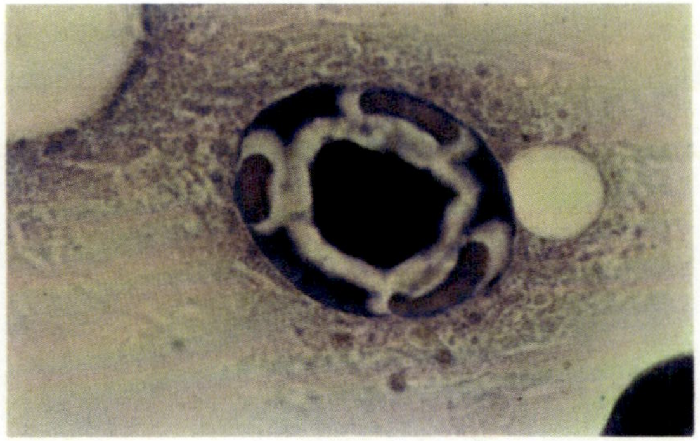

FIGURE 62. Simian adenovirus SV 20 in baboon kidney cell. Late stage, with red bars in nucleus. (H & E; magnification × 800.)

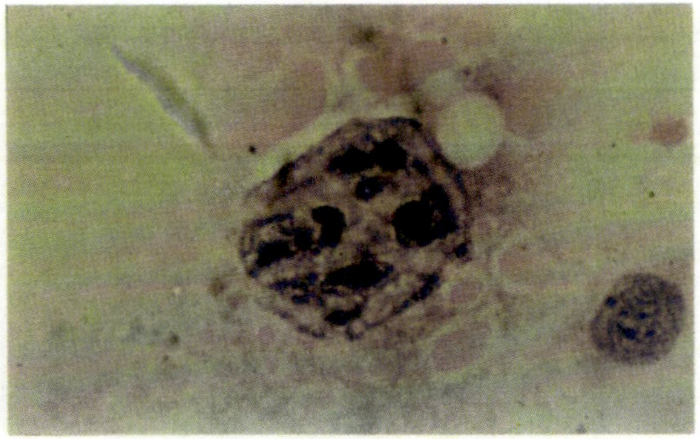

FIGURE 63. Human adenovirus type 5 in human amnion cell. Nuclear crystals, large cytoplasmic inclusions. (H & E; magnification × 800.)

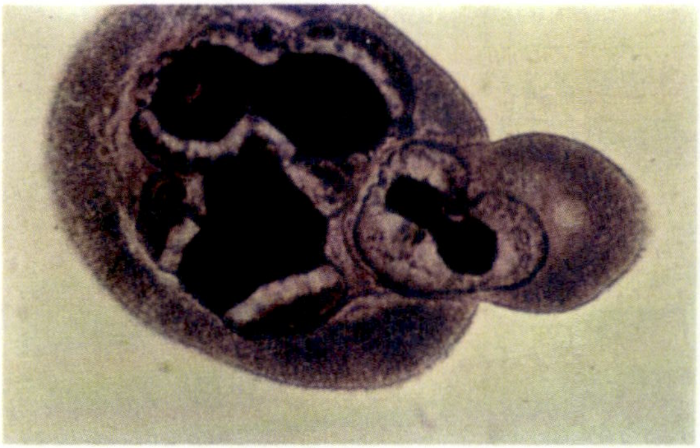

FIGURE 64. Simian adenovirus strain in baboon kidney cells. Cytophagocytosis, with partially ingested cell. (H & E; magnification × 800.)

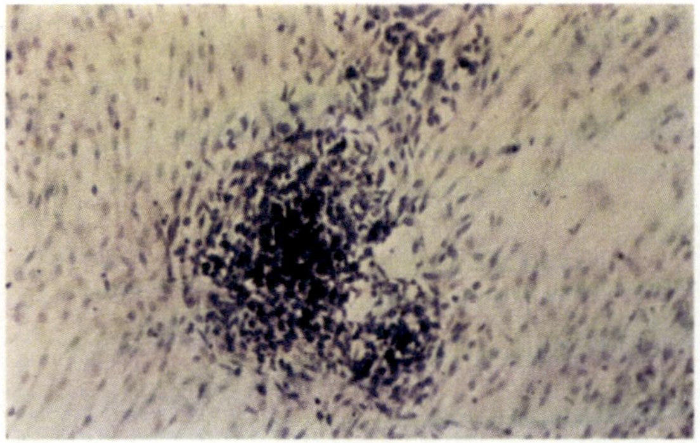

FIGURE 65. Simian adenovirus SA 7 transformation focus in hamster cell culture. (H & E; magnification × 120.)

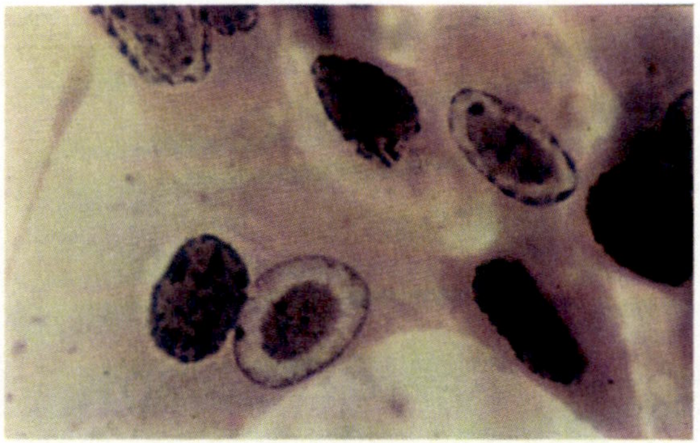

FIGURE 66. Herpes simplex virus in vervet kidney cells. Diffuse basophilic and contracted eosinophilic inclusions. (H & E; magnification × 800.)

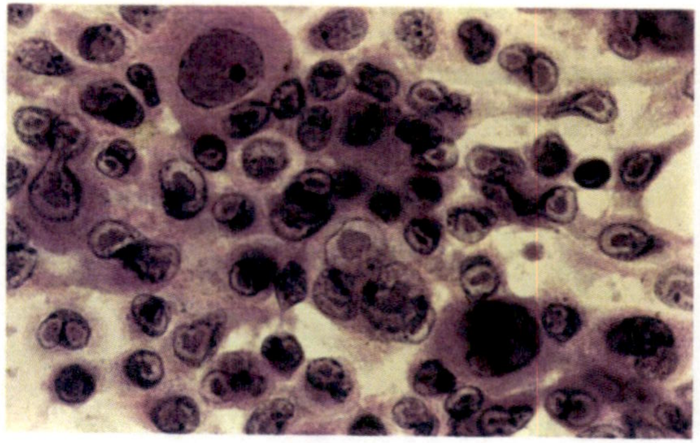

FIGURE 67. Herpes simplex virus in MA 104 rhesus kidney cells. (H & E and phloxine; magnification × 320.)

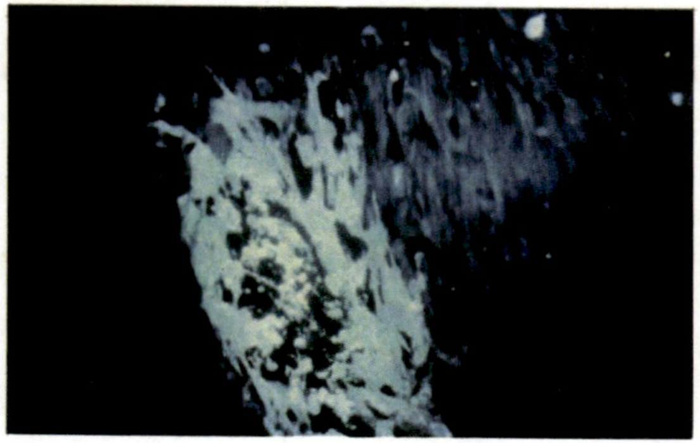

FIGURE 68. Herpes simplex virus in human fetal lung cells. (Fluorescein-stained focus; magnification × 200.)

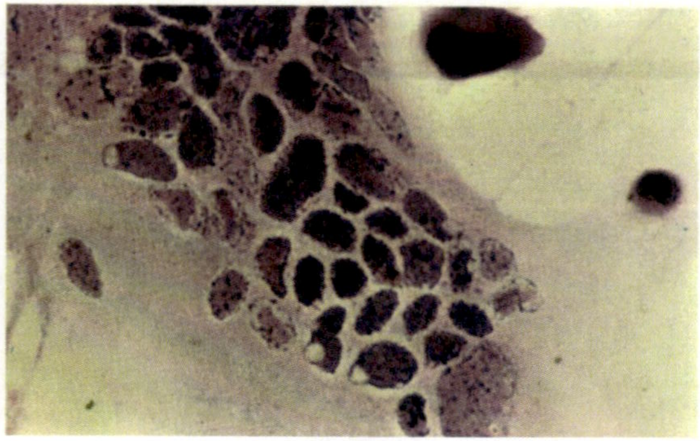

FIGURE 69. Simian herpesvirus SA 8 in vervet kidney cells. (H & E; magnification × 320.)

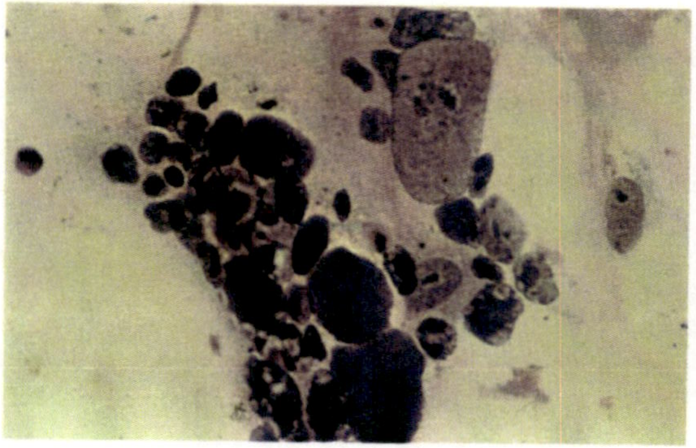

FIGURE 70. Simian herpesvirus SA 15 in vervet kidney cells. Marked nuclear variation. (H & E; magnification × 320.)

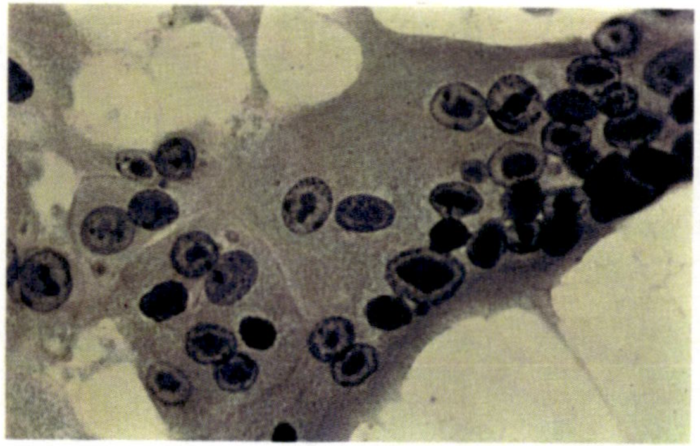

FIGURE 71. Spider monkey herpesvirus in Vero cells. (H & E; magnification × 320.)

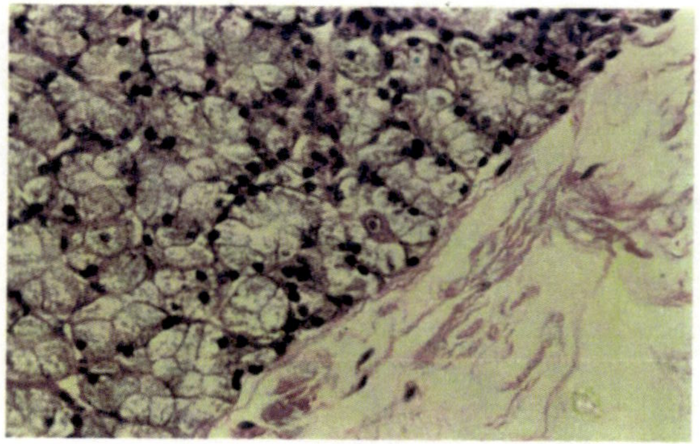

FIGURE 72. Simian cytomegalovirus in vervet monkey kidney parotid gland section. Cf. Figure 73. (H & E; magnification × 200.)

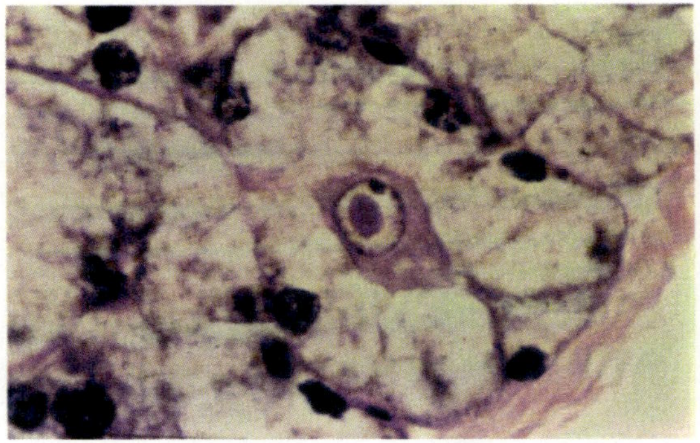

FIGURE 73. Simian cytomegalovirus in vervet parotid gland section. Cf. Figure 72. (H & E; magnification × 800.)

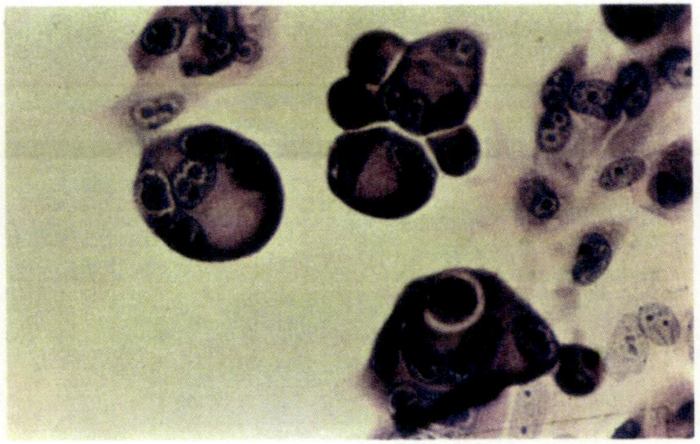

FIGURE 74. Simian cytomegalovirus SA 6 in vervet monkey kidney cells. Cytophagocytosis. (H & E; magnification × 320.)

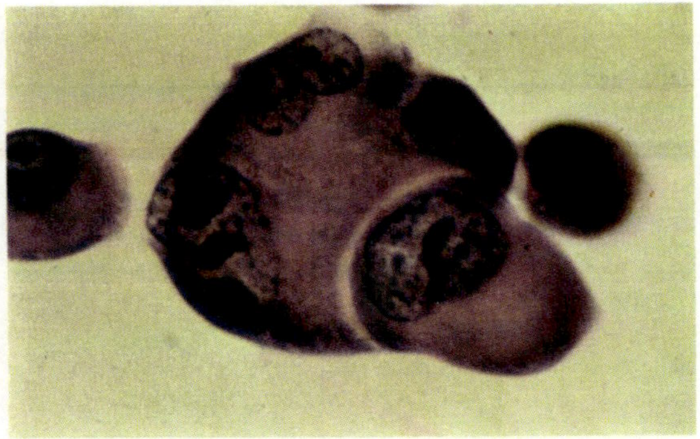

FIGURE 75. Simian cytomegalovirus SA 6 in vervet monkey kidney cells. Cytophagocytosis, with partially ingested cell. (H & E; magnification × 800.)

FIGURE 76. Simian cytomegalovirus in vervet monkey kidney cells. (Acridine orange; magnification × 1000.)

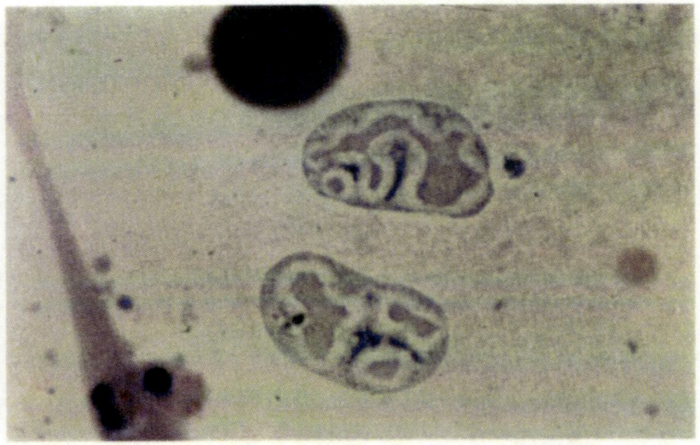

FIGURE 77. Varicella virus in vervet monkey kidney cells. (H & E; magnification × 800.)

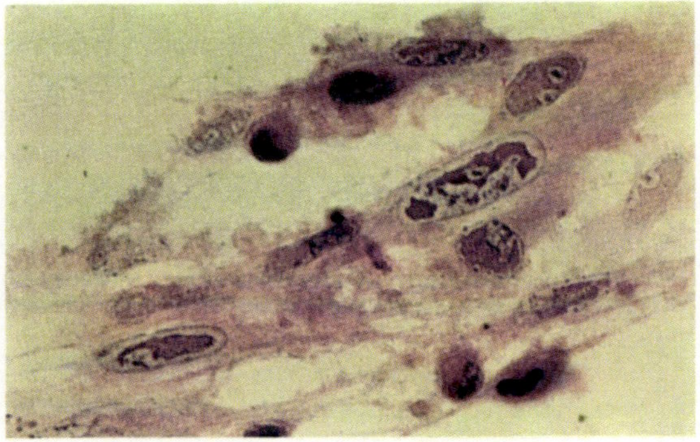

FIGURE 78. Varicella virus in human embryo lung cells. (H & E and phloxine; magnification × 320.)

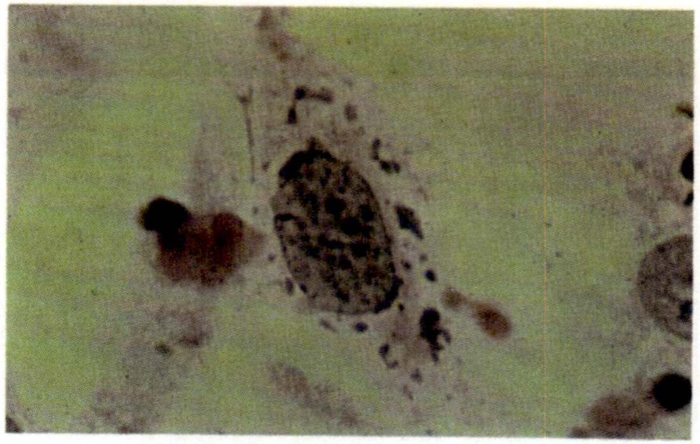

FIGURE 79. Influenza B virus in vervet monkey kidney cell. (H & E; magnification × 800.)

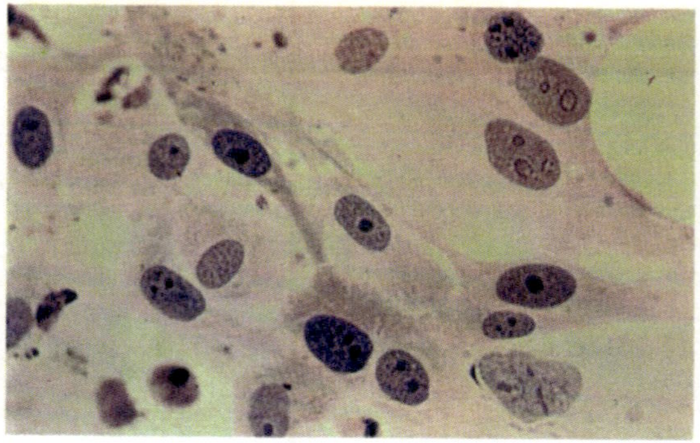

FIGURE 80. Baboon influenza virus in baboon kidney cells. (H & E; magnification × 320.)

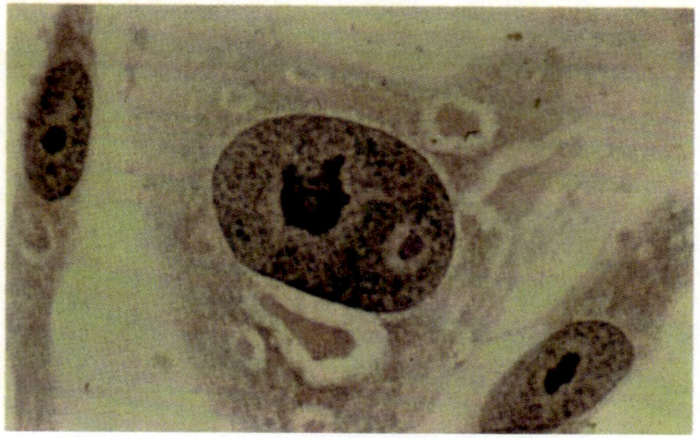

FIGURE 81. Simian parainfluenza virus SA 10 in vervet monkey kidney cell. (H & E; magnification × 800.)

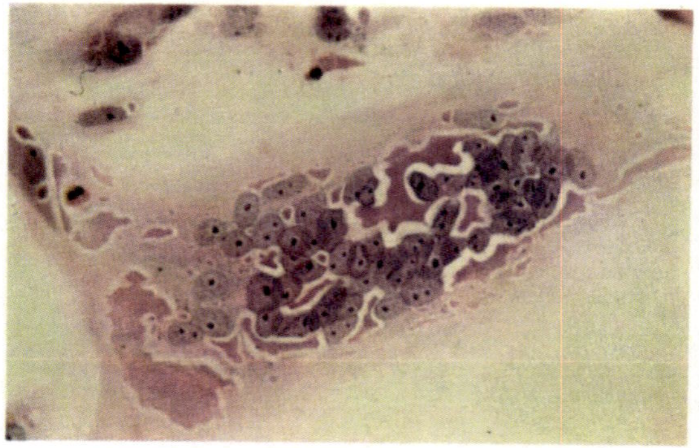

FIGURE 82. Human parainfluenza type 2 virus in vervet monkey kidney cells. (H & E; magnification × 200.)

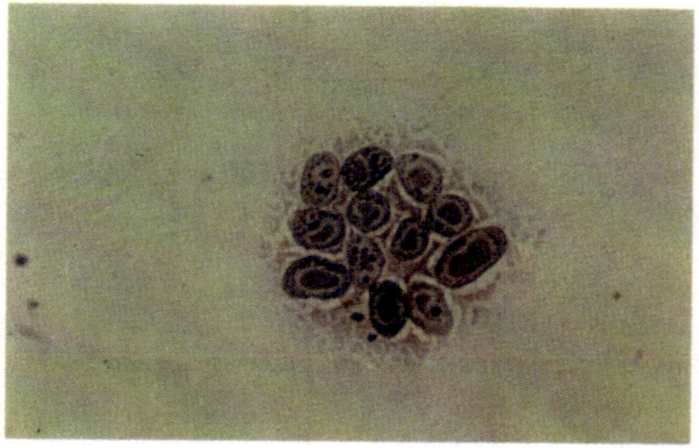

FIGURE 83. Measles virus in vervet monkey kidney cells. (H & E; magnification × 320.)

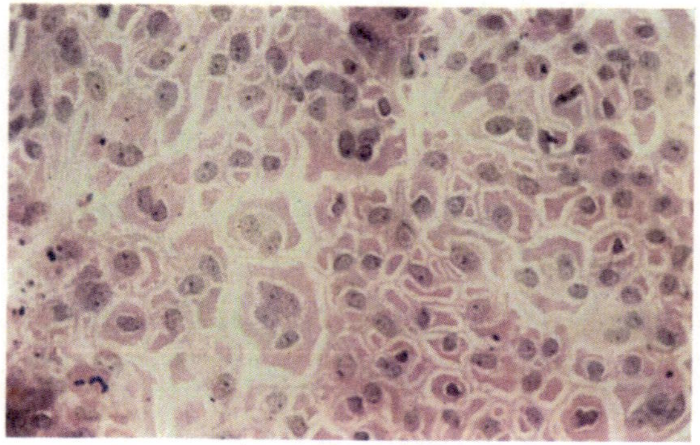

FIGURE 84. Canine distemper virus in vervet monkey kidney cells. Cf. Figure 96. (H & E; magnification × 200.)

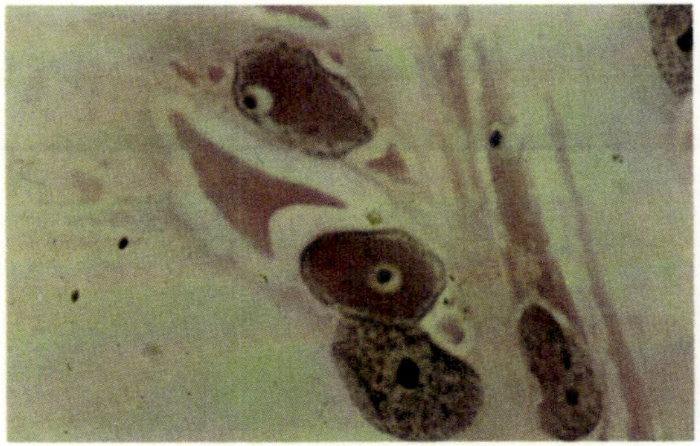

FIGURE 85. Virus from clinical case of mumps in vervet monkey kidney cells. (H & E; magnification × 800.)

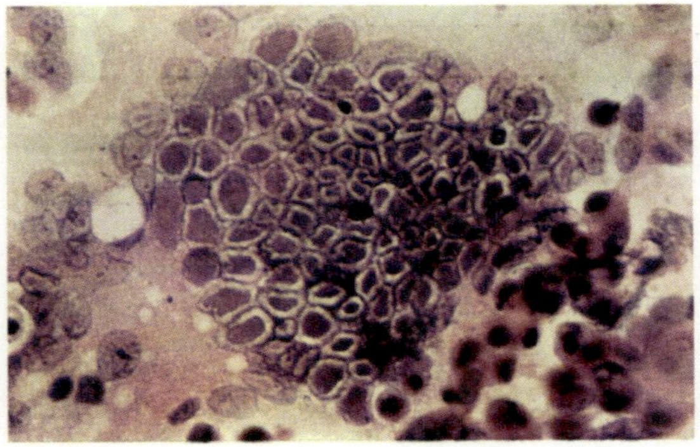

FIGURE 86. Simian spumavirus and human adenovirus in vervet monkey kidney cells. (H & E; magnification × 320.)

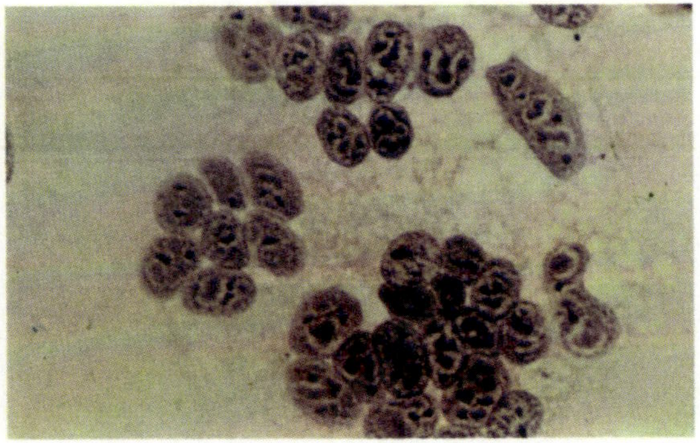

FIGURE 87. Simian spumavirus and SV 40 in vervet monkey kidney cells. (H & E; magnification × 320.)

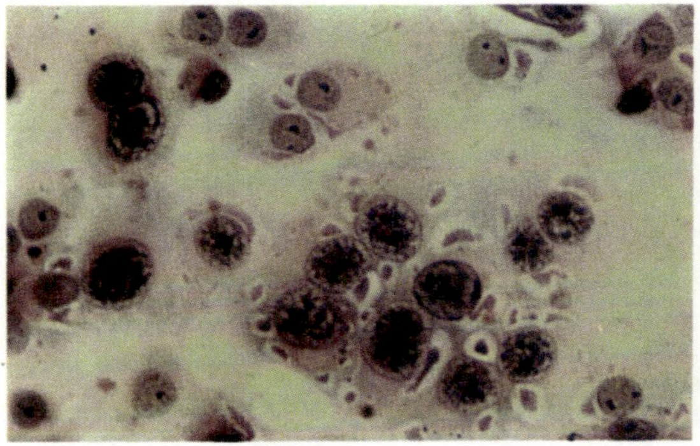

FIGURE 88. SV 40 and rhesus measles virus in vervet monkey kidney cells. (H & E; magnification × 320.)

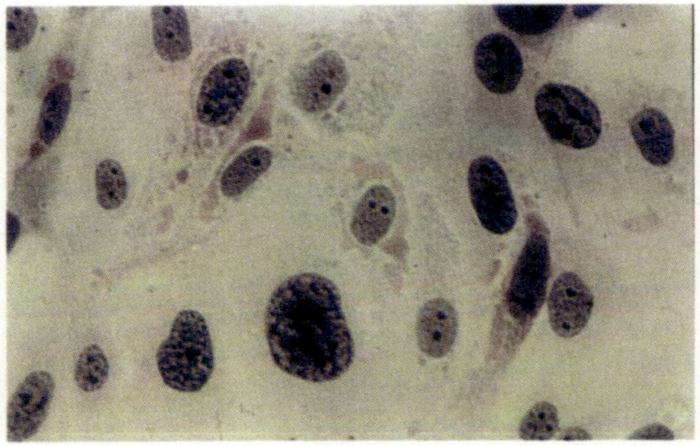

FIGURE 89. SV 40 and reovirus in baboon kidney cells. (H & E; magnification × 320.)

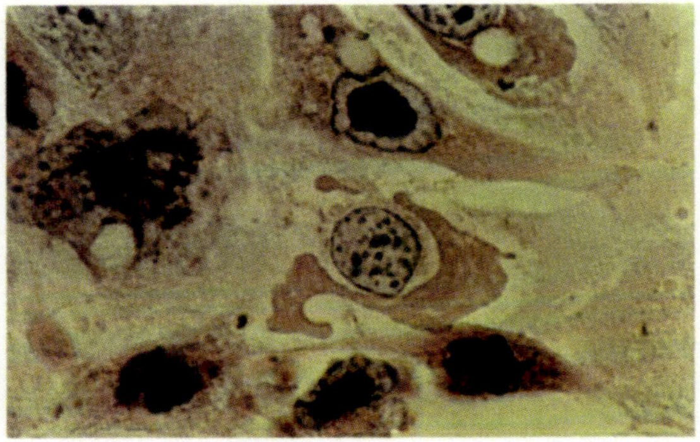

FIGURE 90. Reovirus and adenovirus in vervet monkey kidney cells. (H & E; magnification × 800.)

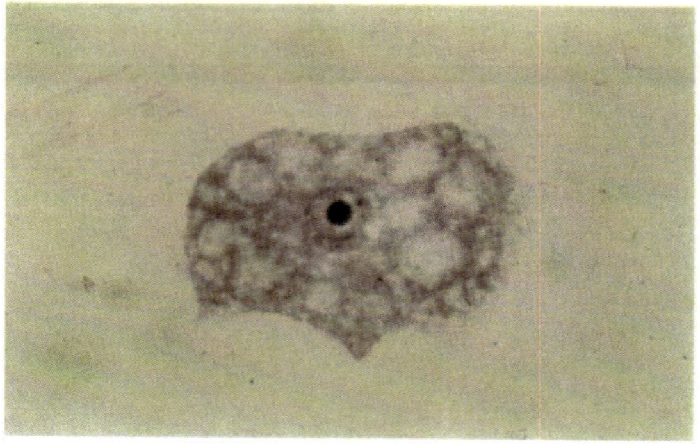

FIGURE 91. Acanthamoeba trophozoite. (H & E; magnification × 1250.)

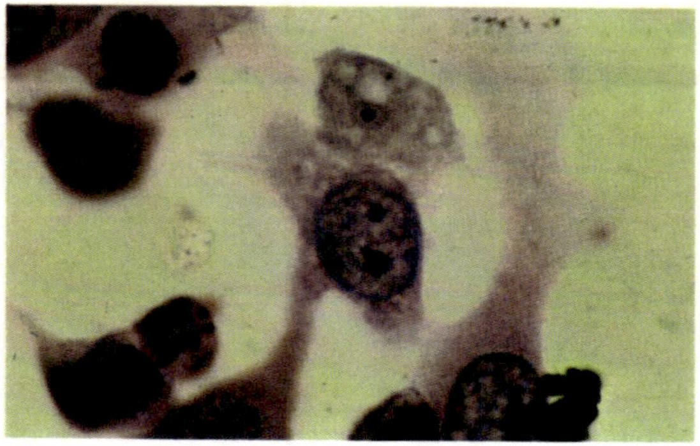

FIGURE 92. Acanthamoeba trophozoite in HeLa cell culture. (H & E; magnification × 800.)

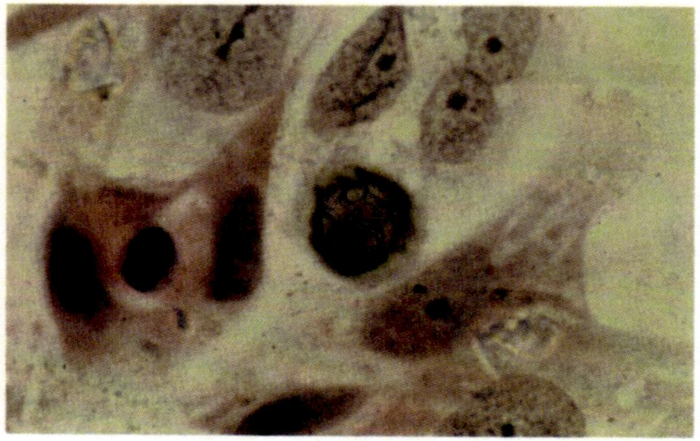

FIGURE 93. Acanthamoeba cyst with pores, in vervet monkey kidney cell culture. (H & E; magnification × 800.)

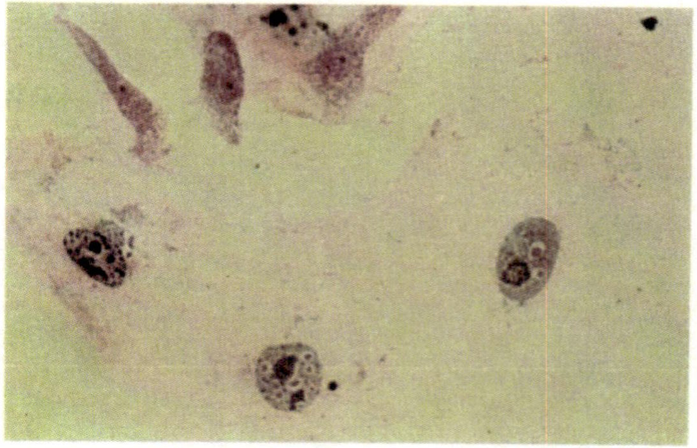

FIGURE 94. Unidentified amoebae in rhesus monkey kidney cell culture. (H & E; magnification × 320.)

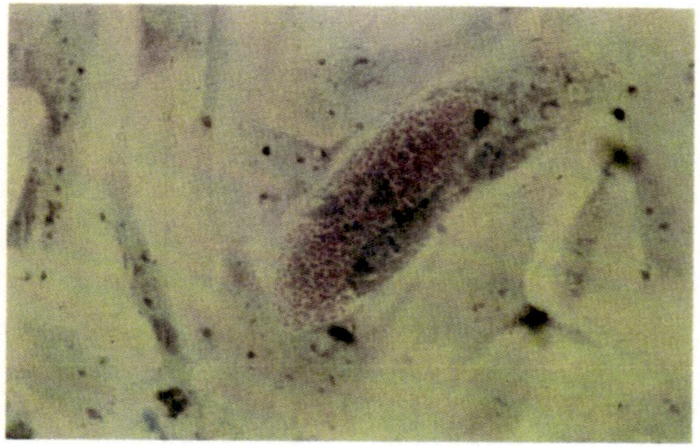

FIGURE 95. *Encephalitozoon cuniculi* in rabbit kidney cell culture. (Wright and Craighead; magnification × 320.)

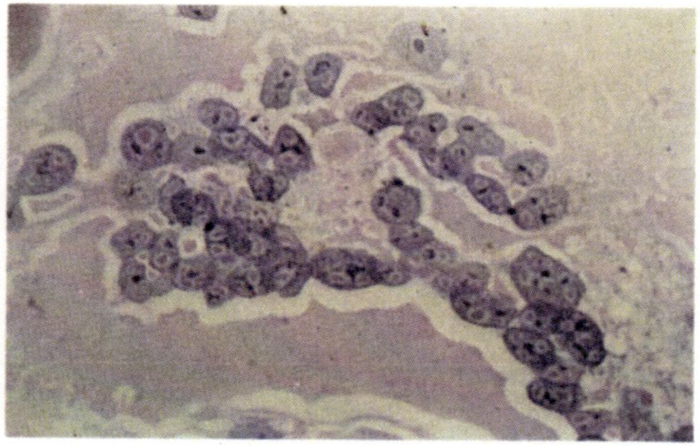

FIGURE 96. Canine distemper virus in vervet monkey kidney cells. Syncytium with inclusions. Cf. Figure 84. (H & E; magnification × 320.)

Chapter 7

CYTOPATHIC CHANGES AFFECTING BOTH CYTOPLASM AND NUCLEUS

I. ORTHOMYXOVIRUSES (INFLUENZA A, B, AND C)

It is generally accepted that the nucleus is essential for the replication of influenza viruses, and enucleated cells will not support their growth.[1,2] Studies with fluorescein- or ferritin-tagged antibodies have shown that the S (soluble) nucleocapsid antigen, which is formed in the cytoplasm, rapidly moves to the nucleus where it may appear in the nucleolus, and then in the rest of the nucleoplasm, before being found again in quantity in the cytoplasm.[3,4] A putative nonstructural viral protein is also associated with the nucleolus.[5] Electron microscopy has shown[6] that the nucleoli of infected cells hypertrophy and lose their compactness before becoming markedly reduced in size. The V antigens, which include hemagglutinin and neuraminidase, are found only in the cytoplasm.

Influenza viruses can be cultivated in the developing chick embryo and also in some cell culture systems of which primary monkey kidney cells are commonly used, particularly for the isolation of influenza B. Influenza C can be propagated in primary chick embryo cells, especially kidney, as well as in primary rhesus monkey kidney cells and the Madin-Darby line of canine kidney cells.[7-9] Influenza C produces little or no cellular destruction, and hemadsorption or hemagglutination with fowl or rat (but not guinea pig or sheep) erythrocytes can be used to detect the presence of virus.[8-10]

Various cell cultures support only abortive infections by influenza viruses, with the production of noninfectious hemagglutinin. Such cultures can be used as indicator systems to demonstrate the presence of virus by hemagglutination or the hemadsorption technique of Vogel and Shelokov et al.[11,12] Immunofluorescent and peroxidase methods have been used to localize viral antigens.[13] Influenza B readily forms plaques under agarose in bovine kidney cell cultures, but influenza A strains plaque better in chick kidney cells.[14] Influenza strains that do not destroy cells may be assayed in chick embryo cell cultures without a gel overlay, by the detection of hemadsorbing plaques 48 hr after infection.[15]

Negroni and Tyrrell[16] studied the morphological changes produced in calf kidney cultures by several influenza A strains which differed in the effects they caused. Rounding with fallout of cells tended to occur particularly with a low concentration of virus inoculum, whereas a high multiplicity of infection led to the survival of many cells. Cytopathic changes were more rapid and extensive when serum was omitted from the medium and when the cultures were rolled or rocked. Cytoplasmic vacuolation occurred in some cells that were not rapidly destroyed, and certain cells had nuclei that became irregularly lobed and finally disintegrated. Mouse and monkey kidney cells were also infected and showed severe cellular damage.

Harford and Hamlin[17] observed basophilic intracytoplasmic bodies in bronchial epithelial cells of irradiated mice inoculated in vivo with influenza A virus. Electron microscopy[18] indicated that these bodies were composed largely of particles corresponding in size to influenza virus. Grossberg[19] noted eosinophilic paranuclear cytoplasmic inclusions in primary hamster kidney cells infected with influenza A and B viruses, but as a few of these inclusions were also present in uninfected cultures their nature remained uncertain. In 1966 Harwin et al.[20] reported changes seen in primary vervet monkey kidney cell cultures infected with influenza B strains and then stained with hematoxylin and eosin. Cytoplasmic inclusions, described as slightly basophilic, appeared first as small irregular bodies that enlarged and coalesced, sometimes surrounding the nucleus. Changes in the nucleus included margination of chromatin and the development of a central eosinophilic mass having strands of chromatin radiating from it. Porebska et al.[21] confirmed that influenza B regularly produced well-defined and strongly basophilic inclusions in primary rhesus monkey kidney cells. Nuclear changes, when present, consisted of margination of chromatin and a mass of eosinophilic material that surrounded the nucleolus and largely filled the nucleus. The cytoplasmic inclusions were first seen at 12 hr and increased up to 24 hr, after which extensive degeneration and cell detachment occurred. Neither S nor V antigens could be demonstrated in the cytoplasmic inclusions by immunofluorescence. Electron microscopy showed that the inclusions were associated with large aggregates of closely packed ribosomelike particles, and did not contain virus.

Figure 79* shows a cell with basophilic cytoplasmic inclusions and early eosinophilic nuclear bodies, in a primary vervet monkey kidney culture infected by us with influenza B virus. Other cells in this culture contained larger nuclear inclusions, and nucleolar changes were also present. The latter were recalled when we encountered similar changes in baboon kidney cell cultures inoculated with throat swab suspensions taken from seven baboons in 1975, leading us to suspect the presence of an influenza virus.[22] The baboon agent was identified as an unusual strain of influenza A,[23] being the first naturally occurring influenza virus in nonhuman primates. Figure 80 illustrates the various stages of nucleolar change produced by this virus; from amphophilic hypertrophy through attenuation and annular degeneration, to complete disappearance. The nucleoplasm at the same time undergoes eosinophilic change. Basophilic cytoplasmic inclusions can be seen in two cells in Figure 80. The nucleolar changes seen in this stained culture correlate with the electron microscopic findings of Anisimova et al. for a recombinant strain of influenza A cited above.[6]

II. PARAMYXOVIRUSES: MEASLES AND PARAINFLUENZA TYPE 3 VIRUSES

A. Measles Virus

Enders and Peebles[24] first isolated the virus in monolayer cultures of human and monkey kidney cells, noting the production of vacuolated syncytia and eosinophilic nuclear inclusions. No mention was made of cytoplasmic inclusions. Reissig et al.[25] cultivated the Edmonston strain of measles virus in HEp-2 cells, and observed that both cytoplasmic and nuclear inclusions were formed. They demonstrated that when the medium contained glutamine the infected cells were predominantly spindle shaped, but in glutamine-deficient medium large syncytia were present. Milovanovic et al.[26] experienced difficulty in adapting to human amnion cells the Edmonston strain after it had been serially passaged in human kidney cells. In amnion cultures the virus induced two cellular morphological changes, namely, syncytia in which nuclear and cytoplasmic inclusions were present, and spindle-shaped or stellate cells, some of which had nuclear inclusions. Continued passage in amnion cells led to predominance of spindle cells over syncytia. Fluid from the 28th amnion passage was titrated simultaneously in human amnion and kidney cells. In the amnion titration only spindle cells were seen, whereas in the kidney cell titration only syncytia were formed. Seligman and Rapp[27] confirmed that passage of the Edmonston strain in human amnion cells led to a change from syncytium formation to spindle cell predominance. They then

* All figures appear following page 58.

cultivated the virus in HEp-2 cells, and noted that in one tube only syncytia appeared, whereas spindle cells were seen in all the other tubes. Serial passage of the virus from the syncytium-containing tube consistently produced only syncytia that were not inhibited by glutamine, and it appeared that syncytium production was a genetic characteristic of this variant of the virus. Oddo et al.[28] modified the technique of subculturing and segregated two strains of measles virus: one producing spindle cells, and the other syncytia.

Electron microscopy of the nuclear and cytoplasmic inclusions has shown that both contain filamentous tubular structures probably consisting of viral nucleoprotein.[29] An immunofluorescence study by Llanes-Rodas and Liu[30] on infected human amnion line cells showed that viral antigen appears in both cytoplasm and nucleus before inclusions are seen by light microscopy. The nuclear inclusions reach their maximal eosinophilic staining capacity when viral antigen and nucleic acid have begun to decline in the nucleus, suggesting that the inclusions consist chiefly of material in which there is little virus, in much the same way as the nuclear inclusion produced by herpes simplex virus is an "intranuclear scar" (Lebrun[31]) or a "residuum" left behind by virus that has already escaped from the cell (Crouse et al.[32]).

Apart from inducing morphological changes in the nuclei of cells that clearly cannot undergo mitosis, measles virus is also capable of causing chromosome breaks and pulverization.[33-35]

Eosinophilic nuclear and cytoplasmic inclusions containing nucleocapsid structures occur in the brain cells of cases of subacute sclerosing panencephalitis (SSPE). Isolation of strains of measles virus responsible for this condition has been achieved by cocultivating infected brain cells with HeLa cells. Two such isolates were shown by Horta-Barbosa et al.[36] to differ from a wild measles strain in several respects, including the capability of the SSPE isolates to replicate with cytopathic changes in HeLa, primary chick embryo and primary canine kidney cells which did not support growth of the wild strain.

Measles virus is occasionally encountered as an endogenous contaminant of monkey kidney cell cultures. The studies of Ruckle-Enders[37] have shown that the majority of monkey strains are identical to human measles virus, but we have noted the existence of a virus of African vervet monkeys that produced similar cytopathic changes yet was not neutralized by measles antiserum.[38] It is probable that measles infection is acquired from human sources by monkeys after capture, and is disseminated during transportation and holding.[39] The clinical disease in monkeys is usually mild, and the macular rash tends to be obscured by the animals' hair.[40] In cell cultures the virus derived from monkeys produces the typical measles triad of syncytia, nuclear, and cytoplasmic inclusions.

Figure 83 illustrates a syncytium in vervet monkey kidney cells infected with human measles virus. The nuclear inclusions start as one or more bodies which enlarge and coalesce, displacing the nucleoli and chromatin peripherally. Cytoplasmic inclusion material is often indistinct and widely distributed, tending to form denser and more discrete bodies near to the nuclei. Such multinucleated cells found in the lymphoid tissue of human measles cases have been called Warthin-Finkeldey cells.

The viruses of canine distemper and of rinderpest of cattle are closely related to measles virus. In Figure 84 are shown prominent inclusions formed in single vervet monkey kidney cells by distemper virus which we originally isolated from primary cultures of kidney cells taken from an aged and asymptomatic dog in which the virus had clearly remained latent. Figure 96 illustrates a syncytium formed by the same virus, the cytoplasmic inclusion material having condensed into very large eosinophilic masses.

B. Parainfluenza Virus Type 3

In 1959 Reisinger et al.[41] reported the isolation of a hemagglutinating virus, designated SF-4, from nasal secretions of calves showing signs of the respiratory disease known as shipping fever. The agent caused focal rounding and degeneration in calf kidney cultures in which both nuclear and cytoplasmic inclusions were observed. Abinanti and Huebner[42] compared the SF-4 virus with the HA-1 strain of human parainfluenza type 3, and found that they were virtually identical by neutralization and hemagglutination-inhibition tests. Further isolations of the bovine agent by Abinanti et al.[43] confirmed the association of the virus with respiratory disease in cattle.

Churchill[44] passaged SF-4 virus in primary monkey kidney and HEp-2 cells, with the production of syncytia containing nuclear and cytoplasmic inclusions, but failed to obtain nuclear inclusions when human parainfluenza 3 was inoculated into HEp-2 cells. Love and Suskind[45] infected monkey kidney and HeLa cells with human parainfluenza 3, and by means of cytochemical stains demonstrated the development of intranuclear bodies containing ribonucleoprotein. It was shown by Kasten and Churchill[46] that the nuclear inclusions produced by the human and bovine viruses were cytochemically essentially the same.

The vervet monkey virus SA 10,[38] which is closely related to human parainfluenza 3, induces small nuclear inclusions as well as larger cytoplasmic bodies. The typical appearance of these inclusions is shown in Figure 81. The infected culture sheet presented a ragged or shredded appearance, and only occasional small syncytia were present. Churchill[47] isolated a strain of parainfluenza 3 from *Erythrocebus patas* monkeys during a severe outbreak of respiratory illness in a monkey holding unit in England. Lung and throat swab specimens were inoculated into primary cultures of patas monkey kidney cells and into HEp-2 human cells. The overall cytopathic effect was one of spindle or stellate cell formation, with a few syncytia. On further passage the virus produced large syncytia. Some isolates caused no grossly visible cytopathic changes when first recovered, and were only detected by hemadsorption with guinea pig erythrocytes. The agent was identified as parainfluenza 3 by the hemagglutination-inhibition test. Serological cross reactions between human, bovine, and patas monkey viruses showed that the human and monkey agents were more closely related to each other than to the bovine strain.

REFERENCES

1. **Follett, E. A., Pringle, C. R., Wunner, W. H., and Skehel, J. J.,** Virus replication in enucleate cells: vesicular stomatitis virus and influenza virus, *J. Virol.*, 13, 394, 1974.
2. **Kelly, D. C., Avery, R. J., and Dimmock, N. J,.** Failure of an influenza virus to initiate infection in enucleate BHK cells, *J. Virol.*, 13, 1155, 1974.
3. **Dulbecco, R. and Ginsberg, H. S.,** Myxoviruses, in *Microbiology,* 2nd ed., Davis, B. D., Dulbecco, R., Eisen, H. N., Ginsberg, H. S., Wood, W. B., and McCarty, M., Eds., Harper & Row, New York, 1973, chap. 56.
4. **Ter Meulen, V. and Love, R.,** Virological, immunochemical, and cytochemical studies of four HeLa cell lines infected with two strains of influenza virus, *J. Virol.*, 1, 626, 1967.
5. **Krug, R. M. and Soeiro, R.,** Studies on the intranuclear localization of influenza virus-specific proteins, *Virology,* 64, 378, 1975.
6. **Anisimova, E., Tuckova, E., and Vonka, V.,** Morphological changes in BHK-21 cells infected with S/N (H_2N_1) influenza virus, *Arch. Gesamte Virusforsch.*, 43, 221, 1973.
7. **O'Callaghan, R. J., Loughlin, M., Labat, D. D., and Howe, C.,** Properties of influenza C virus grown in cell culture, *J. Virol.*, 24, 875, 1977.
8. **Lobodzinska, M.,** Reproduction of influenza C Taylor 1233 virus in various tissue systems, *Arch. Immunol. Ther. Exp.*, 20, 497, 1972.

9. **Mogabgab, W. J.**, Influenza C virus in monkey kidney tissue cultures, *J. Bacteriol.*, 83, 209, 1962.
10. **Chakraverty, P.**, The detection and multiplication of influenza C virus in tissue culture, *J. Gen. Virol.*, 25, 421, 1974.
11. **Vogel, J. and Shelokov, A.**, Adsorption-hemagglutination test for influenza virus in monkey kidney tissue culture, *Science*, 126, 358, 1957.
12. **Shelokov, A., Vogel, J. E., and Chi, L.**, Hemadsorption (adsorption-hemagglutination) test for viral agents in tissue culture with special reference to influenza, *Proc. Soc. Exp. Biol. Med.*, 97, 802, 1958.
13. **Hahon, N., Simpson, J., and Eckert, H. L.**, Assessment of virus infectivity by the immunofluorescent and immunoperoxidase techniques, *J. Clin. Microbiol.*, 1, 324, 1975.
14. **Beare, A. S. and Keast, K. A.**, Influenza virus plaque formation in different species of cell monolayers, *J. Gen. Virol.*, 22, 347, 1974.
15. **Hotchin, J. E., Deibel, R., and Benson, L. M.**, Location of noncytopathic myxovirus plaques by hemadsorption, *Virology*, 10, 275, 1960.
16. **Negroni, G. and Tyrrell, D. A.**, Morphological observations on tissue cultures of epithelial cells infected with influenza-A viruses, *J. Pathol. Bacteriol.*, 77, 497, 1959.
17. **Harford, C. G. and Hamlin, A.**, Effect of influenza virus on cilia and epithelial cells in the bronchi of mice, *J. Exp. Med.*, 95, 173, 1952.
18. **Harford, C. G., Hamlin, A., and Parker, E.**, Electron microscopy of early cytoplasmic changes due to influenza virus, *J. Exp. Med.*, 101, 577, 1955.
19. **Grossberg, S. E.**, Cytocidal assay and propagation of human influenza viruses in cultured primary hamster kidney cells, *Proc. Soc. Exp. Biol. Med.*, 113, 546, 1963.
20. **Harwin, R. M., Joosting, A., and Gear, J. H.**, Inclusions in vervet monkey kidney tissue cultures infected with influenza-B viruses, *Lancet*, 1, 1218, 1966.
21. **Porebska, A., Pereira, H. G., and Armstrong, J. A.**, Cytoplasmic inclusions and nuclear changes in cells infected with influenza-B viruses, *J. Med. Microbiol.*, 1, 145, 1968.
22. **Malherbe, H., Strickland-Cholmley, M., and Smith, G. C.**, Isolation of myxovirus-like agent from baboons, *Lancet*, 2, 41, 1975.
23. **Lief, Florence, S.**, personal communication.
24. **Enders, J. F. and Peebles, T. C.**, Propagation in tissue cultures of cytopathogenic agents from patients with measles, *Proc. Soc. Exp. Biol. Med.*, 86, 277, 1954.
25. **Reissig, M., Black, F. L., and Melnick, J. L.**, Formation of multinucleated giant cells in measles virus infected cultures deprived of glutamine, *Virology*, 2, 836, 1956.
26. **Milovanovic, M. V., Enders, J. F., and Mitus, A.**, Cultivation of measles virus in human amnion cells and in developing chick embryo, *Proc. Soc. Exp. Biol. Med.*, 95, 120, 1957.
27. **Seligman, S. J. and Rapp, F.**, A variant of measles virus in which giant cell formation appears to be genetically determined, *Virology*, 9, 143, 1959.
28. **Oddo, F. G., Flaccomio, R., and Sinatra, A.**, "Giant-cell" and "Strand-forming" cytopathic effect of measles virus lines conditioned by serial propagation with diluted or concentrated inoculum, *Virology*, 13, 550, 1961.
29. **Nakai, T., Shand, F. L., and Howatson, A. F.**, Development of measles virus in vitro, *Virology*, 38, 50, 1969.
30. **Llanes-Rodas, R. and Liu, C.**, A study of measles virus infection in tissue culture cells with particular reference to the development of intranuclear inclusion bodies, *J. Immunol.*, 95, 840, 1965.
31. **Lebrun, J.**, Cellular localization of herpes simplex virus by means of fluorescent antibody, *Virology*, 2, 496, 1956.
32. **Crouse, H. V., Coriell, L. L., Blank, H., and Scott, T. F. McN.**, Cytochemical studies on the intran uclear inclusion of herpes simplex, *J. Immunol.*, 65, 119, 1950.
33. **Stich, H. F. and Yohn, D. S.**, Viruses and chromosomes, *Prog. Med. Virol.*, 12, 78, 1970.
34. **Mustafina, A. N., Grachev, V. P., Sabun, M. A., and Avdeeva, L. I.**, The mutagenic effect of measles virus on cell cultures of different origin, *Acta Virol. Engl. Ed.*, 19, 299, 1975.
35. **Gripenberg, U. and Forbes, C.**, Extensive cytological damage caused by measles in African children, *J. Med. Genet.*, 11, 171, 1974.
36. **Horta-Barbosa, L., Fuccillo, D. A., Hamilton, R., Traub, R., Ley, A., and Sever, J. L.**, Some characteristics of SSPE measles virus, *Proc. Soc. Exp. Biol. Med.*, 134, 17, 1970.
37. **Ruckle-Enders, G.**, Comparative studies of monkey and human measles virus strains, *Am. J. Dis. Child.*, 103, 297, 1962.
38. **Malherbe, H., Harwin, R., and Ulrich, M.**, The cytopathic effects of vervet monkey viruses, *S. Afr. Med. J.*, 37, 407, 1963.
39. **Meyer, H. M., Brooks, B. E., Douglas, R. D., and Rogers, N. G.**, Ecology of measles in monkeys, *Am. J. Dis. Child.*, 103, 307, 1962.
40. **Remfry, J.**, A measles epizootic with 5 deaths in newly-imported rhesus monkeys (*Macaca mulatta*), *Lab. Anim.*, 10, 49, 1976.

41. **Reisinger, R. C., Heddleston, K. L., and Manthei, C. A.**, A myxovirus (SF-4) associated with shipping fever of cattle, *J. Am. Vet. Med. Assoc.*, 135, 147, 1959.
42. **Abinanti, F. R. and Huebner, R. J.**, The serological relationships of strains of parainfluenza 3 virus isolated from humans and cattle with respiratory disease, *Virology*, 8, 391, 1959.
43. **Abinanti, F. R., Byrne, R. J., Watson, R. L., Poelma, L. J., Lucas, F. R., and Huebner, R. J.**, Observations on infections of cattle with myxovirus para-influenza 3, *Am. J. Hyg.*, 71, 52, 1960.
44. **Churchill, A. E.**, Intranuclear inclusions produced by bovine para-influenza, *Nature (London)*, 197, 409, 1963.
45. **Love, R. and Suskind, R. G.**, Cytopathology of parainfluenza type 3 virus infection in HeLa and monkey kidney cells in vitro, *Exp. Cell. Res.*, 24, 521, 1961.
46. **Kasten, F. H. and Churchill, A. E.**, Cytochemistry of cytoplasmic and intranuclear inclusions induced by bovine parainfluenza 3 virus (SF-4) in human cell cultures, *J. Histochem. Cytochem.*, 14, 187, 1966.
47. **Churchill, A. E.**, The isolation of parainfluenza 3 virus from fatal cases of pneumonia in *Erythrocebus patas* monkeys, *Br. J. Exp. Pathol.*, 44, 529, 1963.

Chapter 8

INFECTION BY MORE THAN ONE VIRUS

The detection of multiple infections in cell cultures by means of stained coverslips can obviate much unnecessary labor and expense, yet it is often neglected. The procedure has proved of great value particularly in the isolation of viruses from clinical specimens or from heavily infected materials such as sewage or slaughterhouse effluents.

Individual cells can support the growth of more than one virus: as, for instance, polio- and measles viruses,[1] or measles, SV 40, and poliovirus.[2] We have already mentioned in Chapter 5 the combination of simian foamy virus and enteroviruses such as polio- or echovirus, that produces striking nuclear inclusions not observed with either infection alone. This cytopathic effect that we encountered in the course of vaccine control[3] is illustrated in Figure 17,* which shows a syncytium containing prominent nuclear inclusions resulting from the inoculation of a vaccine strain of poliovirus into a cell culture infected with simian foamy virus, with poliovirus antiserum in the medium. In Figure 18 are seen similar nuclear inclusions in a mixed infection with echovirus type 1 and simian foamy virus. The combination of foamy virus and adenovirus is shown in Figure 86.

Such simultaneous infections can lead to mistakes in interpretation, but in many instances the culture will also contain cells that show only the characteristic changes produced by the individual viruses. We have mentioned that cytoplasmic vacuolation is a less constant feature in SV 40 infection than is the presence of basophilic nuclear inclusions, and the latter can be readily observed when SV 40 occurs in other virus preparations. Figure 89 shows contamination of a standard stock of reovirus by SV 40, the two viruses being revealed singly and in combination in different cells. Figure 87 illustrates simultaneous infection with SV 40 and simian foamy virus, while in Figure 88 is seen the combination of SV 40 and measles virus. Figure 90 shows cells individually infected with reovirus and adenovirus in a culture inoculated with effluent from a slaughterhouse.

REFERENCES

1. **Reissig, M.**, Double infection of cells in culture with measles and poliomyelitis viruses, *Ann. N. Y. Acad. Sci.*, 81, 17, 1959.
2. **Hsiung, G. D., Atoynatan, T., and Gluck, L.**, Multiple virus infection of monkey kidney cells in culture, *Proc. Soc. Exp. Biol. Med.*, 121, 562, 1966.
3. **Malherbe, H.**, Vervet simian agent SA 2, *Lancet*, 2, 101, 1973.

* All figures appear following page 58.

Chapter 9

IDENTIFICATION OF SOME CONTAMINANTS

Cellular cross-contamination between cultures of different types can occur unless strict precautions are taken to prevent it, as mentioned in Chapter 3.I. Mycoplasma contamination of cell cultures is common, and the possible sources include laboratory workers themselves, especially if mouth-pipetting is used, animal serum in the nutrient medium, and clinical specimens inoculated into the cultures. Spread from culture to culture can take place either by aerosol or by direct transfer. Methods for the detection of mycoplasmata have been described by Kenny;[1] and Barile[2] has discussed their cytopathic effects and interactions with viruses.

Virus contaminants may be endogenous, including oncoviruses (see Chapter 5.VI.C.) and other viruses latent in tissues used for the preparation of primary cultures.[3] We have found stained coverslip cultures to be particularly helpful in the detection of simian agents.[4] Sera used in nutrient media occasionally introduce animal viruses[5-7] as well as bacteriophages.[8-10] Trypsin can be a source of parvovirus contamination, as we have mentioned in Chapter 6.II. A method of sterilizing trypsin without destroying its proteolytic activity has been described by Melnick and Wallis.[11]

Parasites encountered rarely in primary cell cultures include *Trypanosoma theileri*[12] and microfilariae, which can be detected by their movement. In primary rabbit kidney cultures the sporozoon *Encephalitozoon cuniculi* may be found.[13] Figure 95* shows this parasite in a primary rabbit kidney culture stained by the method of Wright and Craighead. The mature forms stain red, which would distinguish them from toxoplasma which stains blue. The parasite is always found intracellularly, presenting in its early stages of development as long threads which then become nipped off into the short mature forms. In this figure some of the immature elongated threads, which stain blue, can be seen. Goodpasture's stain can also be used, *E. cuniculi* staining dark blue-black or brown in contrast to toxoplasma which stains yellow. These staining methods have been described by Lillie and Fullmer.[14]

Contamination of cell cultures by amoebae can occur either by way of primary cultivation of animal tissues or by inoculation of clinical specimens. Essential references to such occurrences have been cited by Holmgren.[15] The amoeba most often involved is Acanthamoeba, named for its prickle-like surface projections, but since Naegleria has also been recovered from the nose of a healthy child,[16] this genus may be found more often if sought. Those who wish to enter the turbid waters of taxonomy may refer to the article by Chang[17] on free-living amoebae, for any distinction between Acanthamoeba and Hartmannella organisms.

For virological purposes it is common practice to centrifuge clinical specimen suspensions in order to use the supernatant fluid for isolation of virus. Thus amoebae that may be present are discarded, the baby being thrown out and the bath water retained.

In 1961 Chang[18] reported cytopathic changes in cell cultures caused by a transmissible agent thought to be a virus which was subsequently designated the lipovirus. A number of papers were published about this putative virus, but in 1966 Chang acknowledged[19] the amoebic nature of the agent later identified as Acanthamoeba rhysodes.[17] This protozoan should not be confused with the lipid-containing bacteriophage described by Espejo and Canelo[20] to which the name "Lipovirus p-PM2" was officially given.[21]

* All figures appear following page 58.

In 1966 Pereira et al.[22] reported several isolates of a cytopathic agent, called the Ryan virus, in HeLa and rhesus monkey kidney cultures inoculated with specimens from patients with respiratory illness. It was noted that the agent could not be dissociated from the centrifuged deposit of culture harvests. Small eosinophilic nuclear inclusions were seen in stained cultures. When infected cultures were reexamined by electron microscopy it was observed[23] that the mitochondria in certain atypical cells had tubular cristae characteristic of amoebae, and the agent was identified as belonging to the genus *Hartmannella* (or *Acanthamoeba*). Through the courtesy of Dr. R. S. Chang and Dr. J. A. Armstrong we obtained these amoebae and studied their behavior in cell cultures. Figure 91 illustrates an acanthamoeba trophozoite with its vacuolated cytoplasm and round nucleus containing a prominent central karyosome. In Figure 92 is shown another trophozoite measuring 13 × 21 μm in a cell culture. The cyst of acanthamoeba, seen in Figure 93, measures about 11 μm in diameter and has a wrinkled surface with porelike structures.

On one occasion we encountered an unidentified amoebic contaminant of cell cultures which multiplied in discrete foci measuring 1 to 15 mm in diameter, destroying the cells in those areas. The culture cells at the periphery of the foci showed degenerative changes and eosinophilic nuclear inclusions which resembled those produced by the so-called Ryan virus. We have illustrated these inclusions in Figures 6 and 94, the latter also showing three of the amoebae in close proximity to three culture cells. We have assumed that this cytopathic effect is due to a toxin elaborated by the amoeba. Dunnebacke and Schuster[24, 25] have, however, shown that Naegleria produce a factor present in cell-free lysates which is neither a toxin nor a virus. Cell cultures inoculated with the factor remain morphologically unchanged for about 4 days, but then within a period of some 12 hr there occurs an abrupt and widespread degeneration of cells. The factor is smaller than any known virus or viroid, and can be serially passaged. It has been suggested that this substance may play a role in the pathogenesis of primary amoebic meningoencephalitis caused by Naegleria.

The contaminants discussed above can cause considerable confusion unless they are promptly recognized. The use of stained coverslip cultures can aid significantly in their identification, providing another example of the value of this technique in virological investigations.

REFERENCES

1. **Kenny, G. E.,** Contamination of mammalian cells in culture by mycoplasmata, in *Contamination in Tissue Culture,* Fogh, J., Ed., Academic Press, New York, 1973, chap. 5.
2. **Barile, M. F.,** Mycoplasmal contamination of cell cultures: mycoplasma-virus-cell culture interactions, in *Contamination in Tissue Culture,* Fogh, J., Ed., Academic Press, New York, 1973, chap. 6.
3. **Sanders, F. K.,** The presence of viruses in uninoculated tissue cultures: sources and methods of detection, in *Contamination in Tissue Culture,* Fogh, J., Ed., Academic Press, New York, 1973, chap. 11.
4. **Malherbe, H., Harwin, R., and Ulrich, M.,** The cytopathic effects of vervet monkey viruses, *S. Afr. Med. J.,* 37, 407, 1963.
5. **Kniazeff, A. J.,** Endogenous virus contaminants in fetal bovine serum and their role in tissue culture contamination, in *Contamination in Tissue Culture,* Fogh, J., Ed., Academic Press, New York, 1973, chap. 10.
6. **Nuttall, P. A., Luther, P. D., and Stott, E. J.,** Viral contamination of bovine foetal serum and cell cultures, *Nature (London),* 266, 835, 1977.

7. **Geder, L., Hyman, R. W., Figueroa, M., Oakes, J. E., Iltis, J. P., Dawson, M. S., and Rapp, F.**, Identification of a herpesvirus isolated from cytomegalovirus-transformed human cells, *J. Virol.*, 27, 713, 1978.
8. **Merril, C. R., Friedman, T. B., Attallah, A. F., Geier, M. R., Krell, K., and Yarkin, R.**, Isolation of bacteriophages from commercial sera, *In Vitro*, 8, 91, 1972.
9. **Chu, F. C., Johnson, J. B., Orr, H. C., Probst, P. G., and Petricciani, J. C.**, Bacterial virus contamination of fetal bovine sera, *In Vitro*, 9, 31, 1973.
10. **Moody, E. E., Trousdale, M. D., Jorgensen, J. H., and Shelokov, A.**, Bacteriophages and endotoxin in licensed live-virus vaccines, *J. Infect. Dis.*, 131, 588, 1975.
11. **Melnick, J. L. and Wallis, C.**, Problems related to the use of serum and trypsin in the growth of monkey kidney cells, *Dev. Biol. Stand.*, 37, 77, 1977.
12. **Lundholm, B. D., Storz, J., and McKercher, D. G.**, *Trypanosoma theileri* as a contaminant of tissue origin in cultures of fetal bovine kidney cells in vitro, *Virology*, 8, 394, 1959.
13. **Malherbe, H. and Munday, V.**, *Encephalitozoon cuniculi* infection of laboratory rabbits and mice in South Africa, *J. S. Afr. Vet. Med. Assoc.*, 29, 241, 1958.
14. **Lillie, R. D. and Fullmer, H. M.**, *Histopathologic Technic and Practical Histochemistry*, 4th ed., McGraw-Hill, New York, 1976, 734.
15. **Holmgren, N. B.**, Contamination in tissue culture by parasites, in *Contamination in Tissue Culture*, Fogh, J., Ed., Academic Press, New York, 1973, chap. 8.
16. **Shumaker, J. B., Healy, G. R., English, D., Schultz, M., and Page, F. C.**, *Naegleria gruberi:* isolation from nasal swab of a healthy individual, *Lancet*, 2, 602, 1971.
17. **Chang, S. L.**, Small, free-living amebas: cultivation, quantitation, identification, classification, pathogenesis, and resistance, *Curr. Top. Comp. Pathobiol.*, 1, 201, 1971.
18. **Chang, R. S.**, Properties of a transmissible agent capable of inducing marked DNA degradation and thymine catabolism in a human cell, *Proc. Soc. Exp. Biol. Med.*, 107, 135, 1961.
19. **Chang, R. S., Pan, I., and Rosenau, B. J.**, On the nature of the "lipovirus", *J. Exp. Med.*, 124, 1153, 1966.
20. **Espejo, R. T. and Canelo, E. S.**, Properties of bacteriophage PM 2: a lipid-containing bacterial virus, *Virology*, 34, 738, 1968.
21. **Wildy, P.**, Classification and nomenclature of viruses, *Monogr. Virol.*, 5, 40, 1971.
22. **Pereira, M. S., Marsden, H. B., Corbitt, G., and Tobin, J. O.**, Ryan virus: a possible new human pathogen, *Br. Med. J.*, 1, 130, 1966.
23. **Armstrong, J. A. and Pereira, M. S.**, Identification of "Ryan virus" as an amoeba of the genus *Hartmannella*, *Br. Med. J.*, 1, 212, 1967.
24. **Dunnebacke, T. H. and Schuster, F. L.**, The nature of a cytopathogenic material present in amebae of the genus *Naegleria*, *Am. J. Trop. Med. Hyg.*, 26, 412, 1977.
25. **Dunnebacke, T. H. and Schuster, F. L.**, Cytopathogenic material from amoebae of the genus *Naegleria*, in *Microbiology 1977*, Schlessinger, D., Ed., American Society for Microbiology, Washington, D.C., 1977, p. 583.

APPENDIX

A. Method for staining coverslip cultures with Ehrlich's hematoxylin and eosin.

Steps 1 to 12 are all performed without removing coverslip from tube, except for Step 8 after which the coverslip is replaced in the tube.

1. Fix in Bouin's solution for 1 hr.
2. 70% ethyl alcohol for 12 to 24 hr.
3. Remove caps from tubes; wash with running cold tap water until the yellow is out.
4. Rinse with distilled water; drain tubes well.
5. Ehrlich's hematoxylin for 2 hr.
6. Filter hematoxylin from tubes back into stock bottle.
7. Wash with running cold tap water until tissue is blue.
8. Remove coverslip with forceps and dip rapidly in 0.125% hydrochloric acid in 70% ethyl alcohol, then quickly wash in tap water.
9. Rinse in 70% alcohol.
10. Stain with 1% alcohol-soluble eosin for exactly 2 min.
11. Pour eosin back into the bottle, do not filter.
12. Two rinses in 70% alcohol.
    Two rinses in 95% alcohol.
    Two rinses in 100% alcohol.
    Two rinses in reagent quality xylol.
13. Mount inverted on slide while coverslip is wet with xylol.

B. Formula for Ehrlich's hematoxylin solution.

Hematoxylin powder	2.0	g
Ammonium or potassium alum	3.0	g
Ethyl alcohol, 95%	100	mℓ
Neutral glycerol	100	mℓ
Distilled water	100	mℓ
Glacial acetic acid	10	mℓ
Sodium iodate	0.3	g

Dissolve the hematoxylin in the ethyl alcohol. Add the distilled water. Incorporate the remaining ingredients, adding the sodium iodate last. Plug lightly with cotton, and keep in 37°C incubator for 48 to 96 hr, shaking daily. The stain is ready when the color is dark.

C. Formula for Bouin's fixative.

Saturated aqueous picric acid	75	mℓ
Concentrated formaldehyde	25	mℓ
Glacial acetic acid	5	mℓ

D. Formula for eosin solution.

BDH spirit-soluble eosin C.I. 45386	1.0	g
Ethyl alcohol, 70%	99	mℓ

E. Formula for phloxine-eosin solution, for optional use instead of eosin solution.

Phloxine powder	1.0	g
Ethyl alcohol, 70%	99	mℓ

Mix equal volumes of 1% solution and 1% phloxine solution. Stain for exactly 2 min, as for eosin.

F. Mounting medium.

EUKITT (pronounced oykit)
Made by O. Kindler, West Germany.

INDEX

A

B

C

D

E

F

G

H

I

J

K

L

M

N

O

P

R

S

T

U

V

W

X

Y

Z